Directional Drilling and Deviation Control Technology

Directional Drilling and Deviation Control Technology

Editor

Aditya Tripathi

Directional Drilling and Deviation Control Technology

Edited by **Aditya Tripathi**

Printed in 2017

ISBN: 978-1-68117-354-2

Library of Congress Control Number: 2015941546

© 2016 by

SCITUS Academics LLC,
616, Corporate Way, Suite 2, 4766,
Valley Cottage, NY 10989

www.scitusacademics.com

Contents

Preface

Directional Drilling and Deviation Control Technology examines and explains techniques for development drilling through directional wells. It has been written by operating company field engineers assisted by drilling consultants. Illustrations show how measuring and guidance devices work, and general procedures and recommendations for equipment are given for each deviation method. Intended for drilling engineers and supervisors in charge of development operations involving deviated wells, the book may be profitably consulted by project and design engineers working in well siting.

Editor

Assessment of Potential Location of High Arsenic Contamination Using Fuzzy Overlay and Spatial Anisotropy Approach in Iron Mine Surrounding Area

Thanes Weerasiri[1], Wanpen Wirojanagud[1, 2, 3], and Thares Srisatit[4]

[1]Department of Environmental Engineering, Faculty of Engineering, Khon Kaen University, Khon Kaen 40002, Thailand

[2]Centre of Excellence on Hazardous Substance Management, Bangkok 10330, Thailand

[3]Research Center for Environmental and Hazardous Substance Management, Khon Kaen University, Khon Kaen 40002, Thailand

[4]Department of Environmental Engineering, Faculty of Engineering, Chulalongkorn University, Bangkok 10330, Thailand

ABSTRACT

Fuzzy overlay approach on three raster maps including land slope, soil type, and distance to stream can be used to identify the most potential locations of high arsenic contamination in soils. Verification of high arsenic contamination was made by collection samples and analysis of arsenic content and interpolation surface by spatial anisotropic method. A total of 51 soil samples were collected at the potential contaminated location clarified by fuzzy overlay approach. At each location, soil samples were taken at the depth of 0.00-1.00 m from the surface ground level. Interpolation surface of the analysed arsenic content using spatial anisotropic would verify the potential arsenic contamination location obtained from fuzzy overlay outputs. Both outputs of the spatial surface anisotropic and the fuzzy overlay mapping were significantly spatially conformed. Three contaminated areas with arsenic concentrations of 7.19 ± 2.86, 6.60 ± 3.04, and 4.90 ± 2.67 mg/kg exceeded the arsenic content of 3.9 mg/kg, the maximum concentration level (MCL) for agricultural soils as designated by Office of National Environment Board of Thailand. It is concluded that fuzzy overlay mapping could be employed for identification of potential contamination area with the verification by surface anisotropic approach including intensive sampling and analysis of the substances of interest.

INTRODUCTION

Arsenic contamination to the environment can cause an adverse environmental problem that further impacts health. Exposure to arsenic can result in a variety of human health problems, including various forms of cancer (e.g. skin, lung, and bladder), cardiovascular and peripheral vascular disease, and diabetes. Humans may be exposed to arsenic through inhalation, dermal absorption, and ingestion of food, water, and soil [1, 2]. Arsenic is a naturally occurring element present in both inorganic and organic forms in different environmental and biological samples and its concentrations may be increased by anthropogenic contamination [3]. The major sources of arsenic pollution may be natural process such as dissolution of arsenic containing bedrock/minerals and anthropogenic activities, for example, percolation of water from mines, wood preservatives, agricultural chemicals, and discharge from an uncontrolled mining and metallurgical industry. It is estimated that about 60% of arsenic present in the environment is of anthropogenic origin [4–10].

Regarding arsenic occurrence in nature, it geologically occurs in soil. Besides gold mine activities, arsenic-bearing hydrothermal minerals frequently occur in ores containing copper, iron, tin, nickel, lead, uranium, zinc, cobalt, or platinum [11]. Mine drainage refers to surface water or groundwater becoming contaminated with heavy metals, arsenic, and/or sulfuric acid as the water infiltrates into the mine shafts, pits, coal piles, ores processing structures, and wastes impoundments, such as mine tailings piles and disposal ponds [12, 13]. On January 22, 2001, USEPA published a final arsenic rule in the Federal Register that revised MCL for arsenic from 0.05 mg/L to 0.01 mg/L (10 µg/L) for drinking water [14]. Office of National Environment Board of Thailand (NEB) specified MCL of arsenic in soils as 3.9 mg/kg and 27 mg/kg for agriculture and other usages, respectively [15].

In Thailand there is some evidence of arsenic contamination in the area of Wangsaphung district, Loei province. The line governmental agencies had investigated both surface water and groundwater, revealing that arsenic concentration was less than the MCL of 0.01 mg/L specified by USEPA. However, there are still some findings of sick peoples due to arsenic exposure even though the iron mine had been already closed since 2005. With this suspicion, more extensive investigation in the whole area and in environmental medium such as soils and plants is required. In addition, the site contamination assessment in the catchment where the abandoned iron mine is situated is actually needed.

In order to identify the potential location of high arsenic contamination in the iron mine surrounding area, the major approach included fuzzy overlay mapping in ArcGIS and surface interpolation of the data derived from field sampling and analysis of arsenic content by spatial anisotropy approach.

Consequently, the primary objectives of this research were (1) to identify the most potential locations of high arsenic contaminant and (2) to verify the most potential locations by surface interpolation of the studied arsenic content in such identified area.

Study Area

The study area covered the area surrounding the abandoned iron mine situated at the catchment in Wangsaphung district, Loei province, the northeastern region of Thailand as shown in Figure 1. Most area in this catchment is plateau area with the elevation of about 250–300 meters above mean sea level (mmsl). Iron mine is situated in the east of the catchment at the elevation of 250 mmsl. Within three-kilometer radius of the iron mine, there are four villages, including Ban Na Nong Bong, Ban Huai Phuk, Ban Nam Huai, and Ban Tio Noi, as shown in Figure 2. Based on the iron mine location, Ban Tio Noi is the nearest village and Ban Huai Phuk is the farthest village located 3.5 kilometers in distance at 251 mmsl elevation and located about 4.5 kilometers in distance at 271 mmsl elevation, respectively. Most of land use in the study area is paddy rice field and crop cultivation such as banana, tapioca, nut, and rubber.

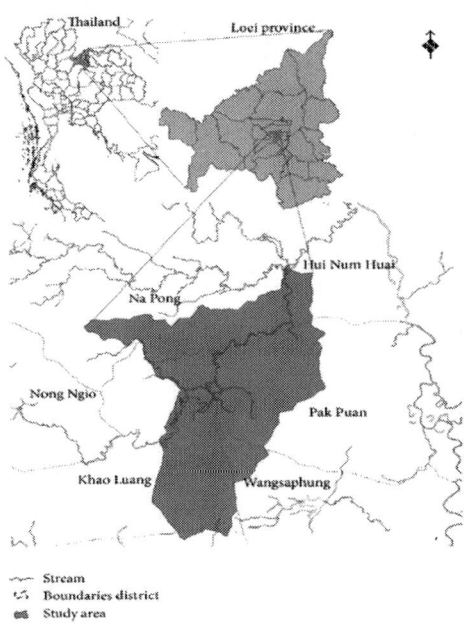

Figure 1: Study area at Wangsaphung district, Loei province, northeast Thailand.

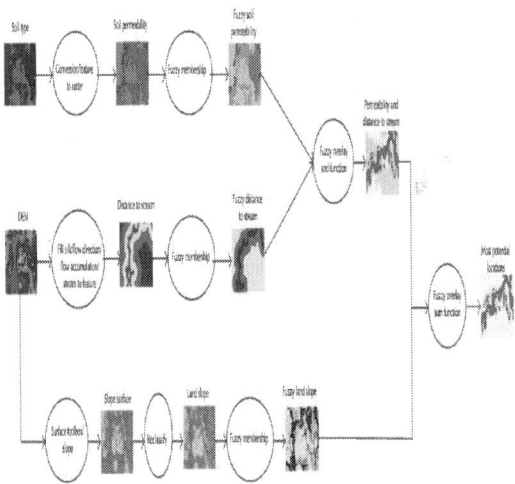

Figure 2: Methodology chart for fuzzy overlay approach.

Within the study catchment there are many small waterways flowing from the high elevation at the top of plateau, about 500–650 mmsl, flowing to the low elevation area and merging to be one stream, namely, Huai stream, passing through the villages downward the Loei River which then joins the Mekong River.

MATERIALS AND METHODS

The conceptual framework of this study is approached with fuzzy overlay mapping to identify the potential arsenic contamination locations and verification of such potential locations with field sampling and analysis of arsenic content in conjunction with the surface interpolation by spatial anisotropy.

Fuzzy Overlay Approach

The source maps selected for fuzzy overlay approach in ArcGIS are based on the factors determining the most potential locations of arsenic contamination which were as follows.

- Land slope: the raster layer of elevation variability is used for slope steepness classification, which affects the rate of lateral movement.
- Soil type: the percentage of sand in soil significantly determines the rate of percolation of water into the groundwater.
- Distance to stream: the raster layer representing the distance from the main stream of each grid cell in the map is used to examine how far the movement is required for the water body.

Raster data of land slope and distance to stream could be created from DEM (digital elevation model) of a cell size of 5 × 5 m. Both layers are continually spatial factors for determining how much water is contaminated by surface runoff process at a site where the stream reaches. The higher slope with shorter distance to stream and sandy soil (higher permeability) make more opportunities for movement of arsenic through the soil pore by surface runoff and then to the stream or store at the plain area along the stream bank. Raster layers of DEM and soil type could be collected from Land Development Department, Ministry of Agriculture and Cooperatives of Thailand. Maps overlay is illustrated in Figure 2.

Sampling and Analysis of Arsenic Content

Soil samples were taken from the potential area derived from the fuzzy overlay mapping. A total of 51 soil samples collection are illustrated in Figure 3. At each location, soil samples were taken at the depth of 0.00–1.00 m from the surface ground level. Methods of drilling and collecting soil samples were performed in accordance with the guidance of American Society for Testing and Materials [16]. Each soil sample was wrapped with aluminum foil sheet and coated with paraffin to protect against the moisture loss and oxidizing reaction that might occur during carrying to the laboratory for analysis. Analysis for temperature, pH, and oxidation-reduction potential (ORP) was made on site. All soil samples were analyzed for arsenic and iron content, OC, CEC, soil type, and its associated parameters such as moisture content, and unit weight. Arsenic contents were analyzed using Inductively coupled plasma mass spectrometry (ICP-MS) method. This technique provides high precision determination of substance, even metallic or nonmetallic, from the relatively small amount of samples [17, 18]. Soil type was classified using mechanical sieve analysis and hydrometer test. Soil group name associated with soil symbol was designated as recommended by Unified Soil Classification System [19].

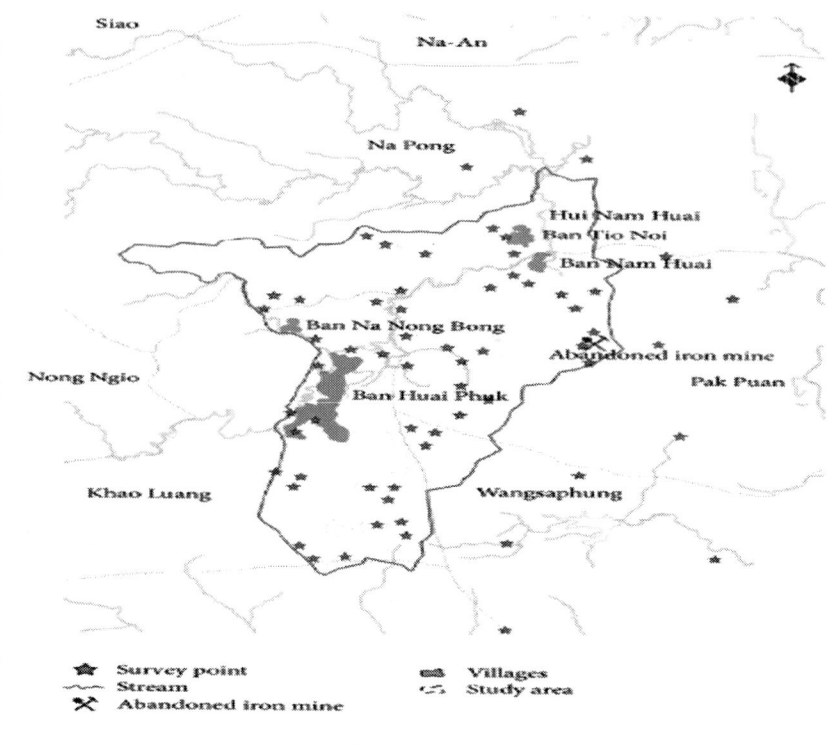

Figure 3: Sampling locations in the study catchment (inside catchment).

Potential Locations Using Fuzzy Overlay

Fuzzy overlay technique employs the principle of fuzzy logic to solve traditional overlay analysis applications in geographic information system (GIS) such as site selection and suitability models. Fuzzy logic is an approach to computing based on "degrees of truth" rather than the usual "true or false" (1 or 0). It is based on the logic of set theory, in which one can traditionally determine whether a value is a member of a set or not. A variation on set theory allows specifying the likelihood that a given value is a member of the set rather than merely specifying whether the value is either in or out of the set [20]. A numeric is used in fuzzy logic with 1 representing full membership in the set and 0 representing nonmembership.

Source layer values using fuzzy overlay approach will be assigned corresponding values on this continuous scale between 0 and 1 according to the likelihood that they have membership in the set. These values are known as "fuzzy membership" values [21].

Fuzzy overlay process composes 4 steps: collecting source layers, assigning fuzzy membership values for each layer, combining the fuzzy layers, and evaluating the results. The result is a layer showing the locations most likely and least likely to be contaminated with pollutants. In this research the source layers are land slope, distance to stream, and soil types specifically concentrated on soil permeability. These three layers are continuous data, called raster maps, of which distance to stream and land slope layers were created from DEM of cell size of 5 × 5 m. Figure 2 presents the procedures of how to establish raster map of distance to stream and land slope from DEM. To create a raster map of soil permeability, the soil type layer was brought as an input raster in the Conversion tool/to Raster/Feature to Raster in toolbox of ArcGIS, the GIS software, also shown in Figure 2.

Assigning Fuzzy Membership

For each source layer, the likelihood that each observed value is a member of the defined set of most potential locations for arsenic contaminant could be specified based on the values for that criterion. Likelihood is indicated by assigning a value on a scale of 1 (very likely) to 0 (not likely). For example, if flat land slope is suitable and steep land slope is not, a value of 1 is assigned to the flat slope and 0 to the steep slope. Between flat and steep slopes, the values are assigned between 1 and 0, accordingly. A new layer, or map, is created corresponding to the fuzzy membership values ranging from 1 to 0. Since there are many observed values in continuous raster layers, it would be more convenient to assign fuzzy membership using mathematical function, which is the relationship between observed values and fuzzy memberships. There are several builtin mathematical functions in ArcGIS such as linear, small, large, MS small, and MS large. Small and large fuzzy function in ArcGIS have been used for assigning fuzzy memberships to the observed values or source values when the relationship between observed values and fuzzy membership is not linear. In this context, fuzzy small and fuzzy large function are used to capture nonlinear relationships between observed values and fuzzy membership values. To assign high fuzzy membership values to small observed values, the small function will be used. Conversely, large function will be used to assign the high fuzzy membership values to large observed values. Following are the assignment details for each source layer in this research.

- Distance to stream: distances to stream were reclassified to be six buffer distances as 100, 500, 1000, 1500, 2000, and 6000 m from the stream and then were assigned for fuzzy memberships with the function small. Parameters inputs in the function small are 500 for the midpoint, 10 for spread, andvery for hedge.

- Land slope: land slopes were reclassified into eight groups from low to high degree of slope as flat to nearly flat, slightly undulating, undulating, rolling, hilly, steep, very steep, and extremely steep,respectively. Fuzzy memberships were then assigned to those classified layers using the function smallwith the input parameters as 5 in degree of slope for midpoint, 10 for spread, and very for hedge.

- Soil permeability: fuzzy memberships were assigned to soil permeability using the function large. Parameters were 3.8 mm/hr, 5, and very for midpoint, spread, and hedge, respectively.

For the small function, the midpoint parameter is the observed value that is assigned for the fuzzy membership value of 0.5, and the spread parameter is the decreasing rate of the fuzzy membership values. The spread defines how tightly the assigned fuzzy membership values cluster around the midpoint. With a large spread value, the fuzzy values decrease rapidly from the midpoint. In case of the large function, midpoint parameter represents the observed value that is assigned for the fuzzy membership value of 0.5, the same as the smallfunction, but the spread parameter controls the rate at which fuzzy membership increases from low to high. Usually, the spread is defined as a value between 1 and 10 [20]. Figure 4 illustrates the fuzzy function small and fuzzy large and their parameters used for assigning fuzzy memberships to each source layer.

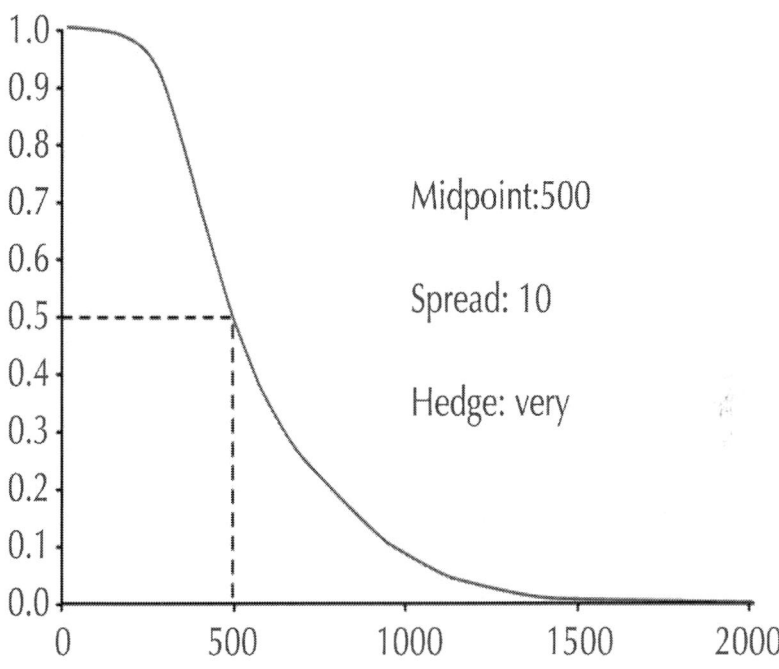

(a) Small function assignments to distance to stream

(a)

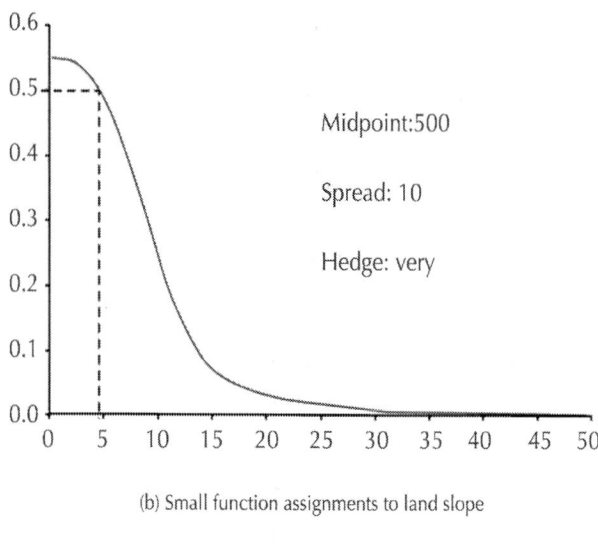

(b) Small function assignments to land slope

(b)

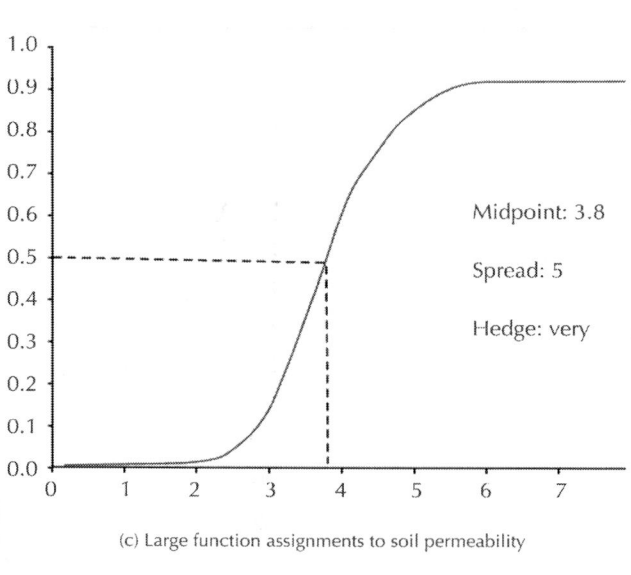

(c) Large function assignments to soil permeability

(c)

Figure 4: Parameters input for fuzzy small and large for assigning fuzzy memberships.

Combining Fuzzy Membership Layers

The overall likelihood of membership in each of the defined sets can be created by combining all fuzzy membership layers, called fuzzy overlay. Fuzzy operators used for combining membership layers in this research are fuzzy AND and Fuzzy SUM. With the AND operator, the output cell will represent the minimum value from all the input fuzzy membership layers, whereas the output values of SUM operator represent the membership value of a cell in each layer as the results of subtracting fuzzy membership of each observed map from 1, followed by multiplying them and then subtracting from 1.

To accomplish the fuzzy overlay in this study, fuzzy AND was first utilized to combine the fuzzy membership layer of soil permeability and distance to stream, and then that result was combined with fuzzy membership layer of land slope using fuzzy SUM, as the process shown in Figure 3.

For this study, the following values are used subsequently.

- In The distance to stream layer and land slope layer, the small function gives the high fuzzy membership values to the small observed values. By applying the input values of 500 as midpoint, 10 as spread, and very for hedge, the fuzzy small function assigned high fuzzy membership values to the distance to stream that is far from stream less than 500 meters. The decreasing rate of fuzzy membership values is controlled by the "spread." If we take the spread parameter with high value in fuzzy small, the fuzzy values will steeply decrease when lower than midpoint.

- For land slope layer, the input parameters are of 5 as Midpoint, 10 as Spread and "very" for Hedge. Fuzzy small function gave high fuzzy membership values for land slope that are less than 5 degrees which cover slope classes of flat to nearly flat and slightly undulating. The other types of slope have been assigned rapidly decreased fuzzy values due to large spread input.

- In the soil permeability layer, the fuzzy large function was used because more permeable capacity of soil would get more chances for arsenic transportation and the relationships between permeability values and the fuzzy membership values are not exactly linear. The midpoint of 1.3 exhibits that the fuzzy membership of 0.5 would be assigned to permeability of 1.3 mm/hr and greater value of fuzzy memberships would be assigned for the higher permeability. The spread of 5 controls the rate at which fuzzy membership increases gradually from low to high.

- The input values for midpoint and spread as described above are based on the expert knowledge, followed by iterative process to modify some fuzzy membership values of the source layers to achieve the best results.

Figure 5 exhibits the layer maps of assigned membership of each data source and next process of their combinations through the use of fuzzy overlay. Fuzzy overlay result of fuzzy membership layers of soil permeability and distance to stream utilizing the AND operator gives the specific areas that are most likely close to stream and associated with the soil condition of high permeability. However, some locations in that result are of high elevation compared to the elevation of the abandoned iron mine which is anticipated as being anthropogenic source of arsenic contaminant. To obtain more reliable result, the fuzzy membership of land slope was overlain with the foregoing result through the fuzzy SUM.

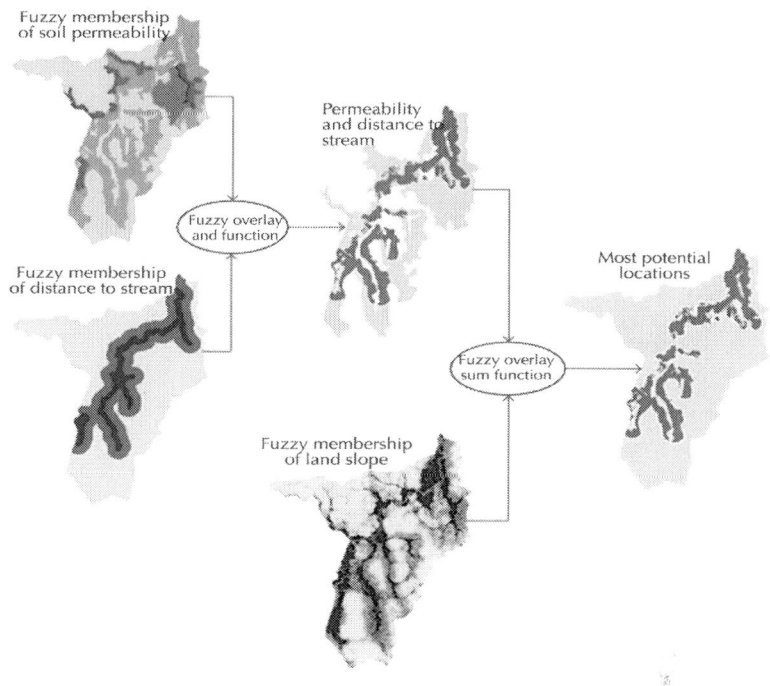

Figure 5: Fuzzy membership layers and process of overlay.

Spatial Anisotropy Assessment

To assess the level of contamination, arsenic concentrations in soils of 51 points in the study area were brought to create surface interpolation using geostatistical techniques, namely, spatial anisotropy. These techniques can not only give a prediction surface but also provide measure of uncertainty or accuracy of the prediction so that they are more reliable than the deterministic technique. In general, the process of establishing surface interpolation is composed of calculating the empirical semivariograms fitting a model and makes a prediction. Kriging, synonymous with "geostatistics" in spatial statistics, is based on the assumption that things that are close to one another and are more alike than those farther away; it generally is called as a spatial autocorrelation. The empirical semivariogram is a mean to explore this relationship.

Pairs that are close in distance should have a smaller measurement difference than those farther away from one another [21, 22]. The extent that this assumption is true can be examined in the empirical semivariogram. Semivariogram can be computed for all pairs of locations separated by distance h, as follows:

Semivariogram (distance h)

$$= 0.5 * \text{average} \left[(\text{value at location } i \right. \quad (1)$$

$$\left. - \text{value at location } j)^2 \right].$$

(1)

There are several models that can be used to fit the empirical semivariogram. Those are, for example, circular, spherical, tetraspherical, exponential, Gaussian, and so forth. Which type is the best model can be determined from the following diagnostics.

- The mean prediction error should be near zero to assure that the prediction errors are unbiased. Also, the standardized prediction errors, that is, the prediction error divided by prediction standard errors, should be near zero.
- The root-mean-square prediction errors should be small. The smaller the root-mean-square the better prediction error.
- The average standard errors should be close to the root-mean-square prediction errors. If the average standard errors are greater than the root-mean-square prediction errors, then there is overestimating in the variability of the predictions; if the average standard errors are less than the root-mean-square prediction errors, then there is underestimating in the variability of the predictions. Another way is to check the root-mean-square is that standardized error should be close to 1.

Crossvalidation chart in ArcGIS can give these diagnostics for testing how well the model predicts the values at unknown locations [22].

For the field data of arsenic concentration and the assumption of exactly unknown but constant mean, the ordinary kriging, one of the geostatistical methods, is chosen to predict the values at the unmeasured points. By examining semivariogram in any directions, it was found that the shape of semivariogram curve varies more with direction, so that the anisotropic approach has been selected as an application in ordinary kriging. In addition, the spherical type of fitting model for semivariogram is appropriate through testing the diagnostics described above.

RESULTS AND DISCUSSION

Based on fuzzy overlay approach, the results shown in Figure 6 exhibit the potential areas of the most contaminated location with arsenic. Such areas demonstrate more chances of capturing, accumulating, or being the pathway of any polluted materials transported by water because the locations are low in elevation with flat slope, high soil permeability, and distance adjacent to the stream. There are three areas being the most potentially contaminated with the high concentration of arsenic. The first area is adjacent to Ban Huai Phuk, 4.5 kilometers in the west of the iron mine. The second area is at the border of the deciduous forest, 2.8 kilometers in the northwest of the iron mine. The third area is Ban Tio Noi and Ban Nam Huai, 3.5 kilometers in the north of the iron mine. It thus preliminarily indicated Ban Huai Phuk, Ban Tio Noi, and Ban Nam Huai being as threatened by high level of arsenic. In order to ascertain the fuzzy overlay results, this study had investigated arsenic concentration in soil for 51 sapling points within the study catchment (see Figure 3). Surface interpolation by spatial anisotropy approach was employed for verification of the Fuzzy Overlay outputs.

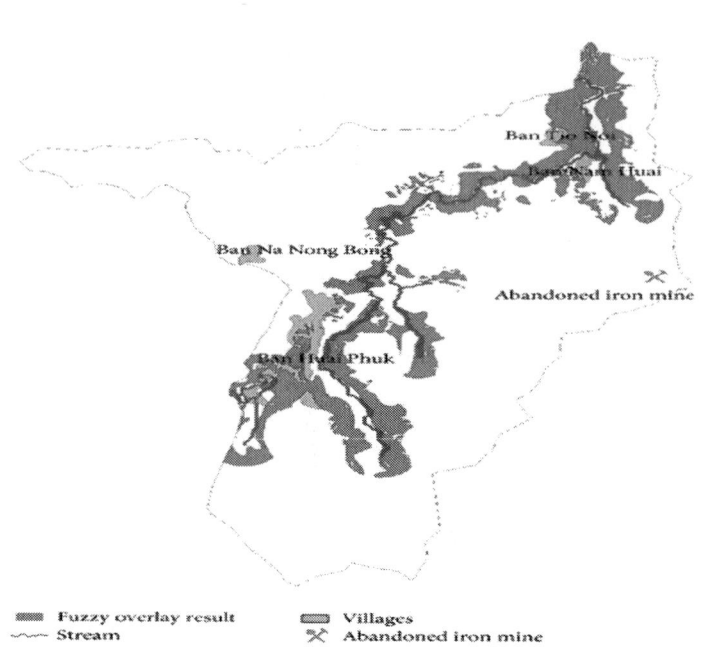

Figure 6: Fuzzy overlay result exhibiting the most potential area of high arsenic content.

Based on descriptive statistics, arsenic concentration in soils within study catchment was 4.75 ± 4.59 mg/kg as stated in Table 1.

Table 1: Quantity of arsenic concentration in soils at the inside and outside of the study catchment, other parameters, and their basic statistics

	Depth (m)	Measured range (mean value ± SD)	Median	Skewness	Kurtosis
As (mg/kg)	0.00	0.29–18.48 (4.75±4.59),	3.41	1.67	1.93
Fe (mg/kg)	0.00	2800–38500 (16300±8396),	14500	0.42	−0.53
pH	0.00	4.20–8.53 (7.44±0.80),	7.67	−2.48	4.27
Temp (°C)	0.00	23.90–39.50 (29.70±3.62),	29.10	0.96	0.69
ORP (mV)	0.00	−104.80–413.70 (225.78±106.85),	248.80	−1.14	1.97

Surface interpolation of arsenic content in soils was produced using ordinary kriging. After detrending and taking the log transformation to arsenic data, empirical semivariogram model was then built from semivariogram scatter plots. Type of spherical anisotropic model was selected as fitted curves for semivariogram models because arsenic data changed not only with the distance but also with the direction. Through the use of MATLAB, three-dimensional anisotropic models of semivariogram could be built, which illustrate semivariogram curves and their parameters in all directions [23]. For example, one semivariogram model of arsenic dataset, shown in Figure 7, presents the values of 1.124 for sill, 1380.49 m for minor range, 2814.20 m for major range, and 38.1° due north for directional angle. Crossvalidation, which helps make an informed decision as to which model provides the best predictions, exhibits the value of mean prediction errors, mean standardized prediction errors, root-mean-square prediction errors, and average standard errors as 0.2485, −0.02989, 3.526, and 3.416, respectively. Consequently, the surface interpolation obtained gives reliable prediction of arsenic content in soils at the unmeasured points.

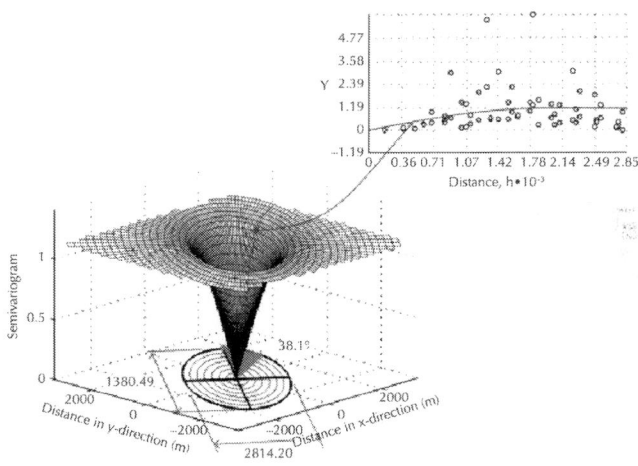

Figure 7: Semivariogram in spatial anisotropy. The top right represents the fitted curve of semivariogram in that direction.

Figure 8 illustrates the results of surface interpolation using ordinary kriging as previously described. It contains 3 maps including prediction map, prediction standard error map, and probability map. The prediction map, in Figure 8(a), presents four local zones of high arsenic content. Placing the prediction map on the fuzzy overlay result, as illustrated in Figure 9, it is obviously seen that, except the zone of iron mine, the other three zones of high arsenic content match the areas in which they were found by the fuzzy overlay approach. Evidently, the fuzzy overlay approach can prove being the most effective way for providing preliminary information necessary for further field works.

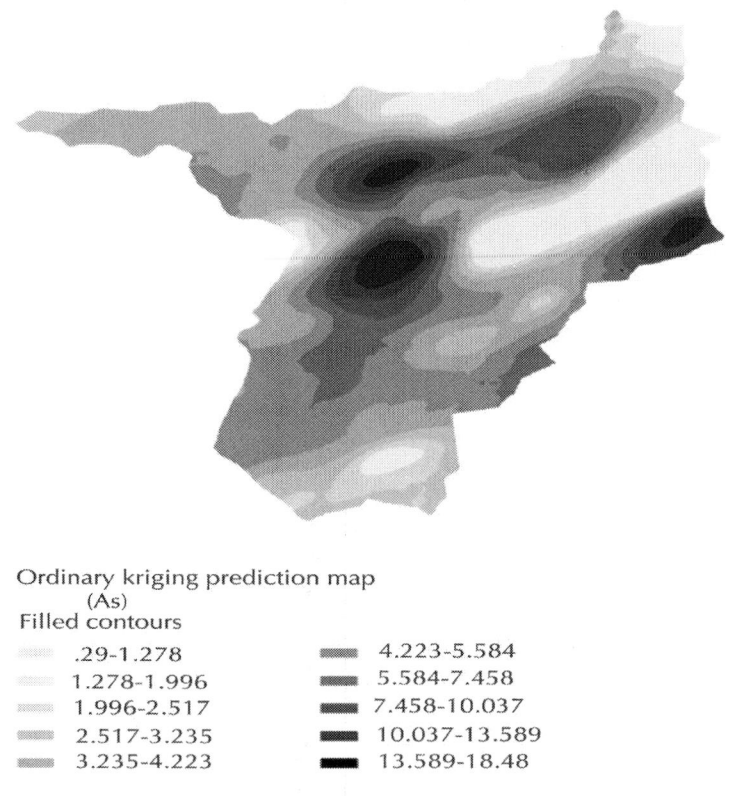

Ordinary kriging prediction map
(As)
Filled contours

.29-1.278		4.223-5.584
1.278-1.996		5.584-7.458
1.996-2.517		7.458-10.037
2.517-3.235		10.037-13.589
3.235-4.223		13.589-18.48

Figure 8: Surface interpolation obtained from spatial anisotropy using ordinary kriging.

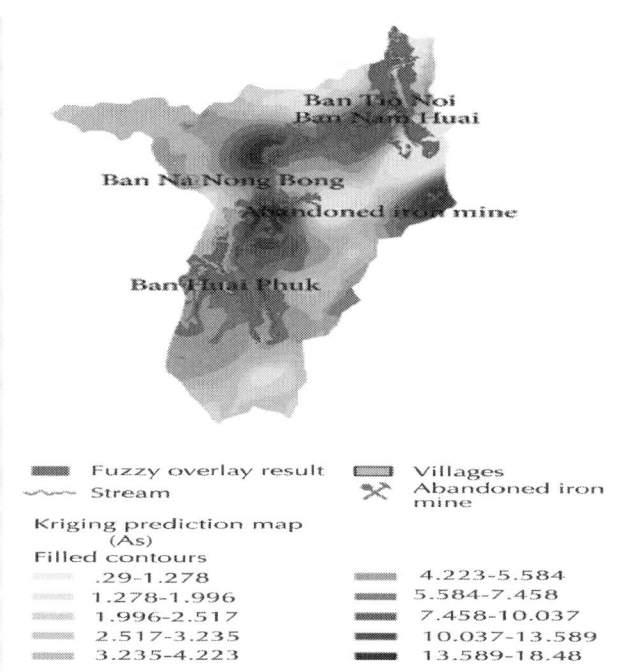

Legend:
- ▬ Fuzzy overlay result
- ∿ Stream
- ▭ Villages
- ⚒ Abandoned iron mine

Kriging prediction map (As)
Filled contours
- .29-1.278
- 1.278-1.996
- 1.996-2.517
- 2.517-3.235
- 3.235-4.223
- 4.223-5.584
- 5.584-7.458
- 7.458-10.037
- 10.037-13.589
- 13.589-18.48

Figure 9: Result of fuzzy overlay match on surface interpolation.

In case of assessment, only prediction model might not be adequate to indicate the contaminant area precisely, and a decision about the classification of safe and unsafe areas with predictions map alone could be inaccurate. To assure the prediction results, prediction standard error and probability maps have been taken into account in considering highly contaminated areas. Prediction standard error in dark-colored areas as shown in Figure8(b) at bottom right is large, indicating much larger variability in prediction. In this case, the usefulness of probability map in Figure 8(c) helps assure the most contaminated areas obtained from the prediction map. The probability map presents the probability of each area that arsenic could exceed the threshold value of 3.9 mg/kg, MCL for agricultural use.

As illustrated in the prediction map, in Figure 8(a), Ban Tio Noi, Ban Nam Huai, and northeast area of Ban Huai Phuk are in the areas of high arsenic content. Also, the probability map, in Figure

8(c), presents 81.5–100%, ensuring the most contaminated arsenic area is at those three villages. Arsenic content in soils at such areas generally exceeded 3.9 mg/kg, the MCL for agricultural use. Within the buffer zone of 500 m from each village, quantity of arsenic in soils and plants is grouped as presented in Table 2. Ban Nam Huai seems to expose more serious risk than other villages because of higher mean value and smaller standard deviation. In paddy rice field, both soil and rice in these three areas should be further intensively investigated as rice is the pathway media to human consumption.

Table 2: Arsenic content in soils within buffer zone of 500 meters from each village (mg/kg)

Villages	Measured range (mean value ± SD)	Plant
Huai Phuk	4.05–12.04 (6.60±3.04)	Rice paddy, corn, longan, mix orchard, mix field crop
Na Nong Bong	2.11–4.64 (3.62±0.80)	Rice paddy, mix field crop, dense deciduous forest
Nam Huai	2.93–10.79 (7.19±2.86)	Rice paddy, corn, mix orchard
Tio Noi	1.95–9.23 (4.90±2.67)	Rice paddy, mix orchard

CONCLUSIONS

Fuzzy overlay is an effective method to preliminarily identify the most potential areas contaminated with arsenic. With the data sources of the distance to stream, land slope, and soil permeability, fuzzy overlay, which is the application in ArcGIS, could give the contaminated locations. In this context fuzzy function and the functions small and large were used to assign fuzzy membership and then we used fuzzy AND and fuzzy SUMto combine layers of fuzzy membership. The function Small was applied for the distance to stream and land slope as the low source values were the most suitable. The function large was applied for soil permeability because the large source values were the most suitable. Result of the fuzzy overlay displayed three locations of high potential arsenic contamination. Accordingly, those locations were in conformance with the area of high arsenic content derived by the spatial anisotropy assessment.

The fuzzy overlay has thus been proved to be straightforward approach for finding the preliminary site study. In general, this method could be applied for any heavy metals or any polluted materials taken with water passing through soil media.

Spatial anisotropy assessment exhibited four zones of high arsenic content in soils. Regardless of iron mine zone which can play major role as an anthropogenic source, the other three zones cover three villages including Ban Huai Phuk, Ban Nam Huai, Ban Tio Noi. Ban Nam Huai is the most risky area because of its high mean value of arsenic contaminant and narrow range of standard deviation. Arsenic content in soils at Ban Nam Huai is 7.19 ± 2.86, exceeding the MCL of 3.9 mg/kg for agricultural soils. Ban Huai Phuk and Ban Tio Noi have also high values and exceeding MCL, but are lower than Ban Nam Huai. The line government agencies should place effective preventive measures and appropriate remediation at this high contaminated area, particularly those three mentioned villages.

ACKNOWLEDGMENT

This work has been supported by the Higher Education Research Promotion and National Research University Project of Thailand, Office of the Higher Education Commission.

REFERENCES

1.   K. R. Henke, Arsenic Environmental Chemistry, Health Threats and Waste Treatment, 2009.

2.   H. K. Chopra and A. Parmar, Engineering Chemistry: A Textbook, Alpha Science, Mumbai, India, 2007.

3.   M. C. Villa-Lojo, E. Alonso-Rodr´ıguez, P. Lopez-Mah ´ ´ıa, S. Muniategui-Lorenzo, and D. Prada-Rodr´ıguez, "Coupled high performance liquid chromatography-microwave digestionhydride generation-atomic absorption spectrometry for inorganic and organic arsenic speciation in fish tissue," Talanta, vol. 57, no. 4, pp. 741–750, 2002.

4.   K. Loska, D. Wiechula, B. Barska, E. Cebula, and A. Chojnecka, "Assessment of arsenic enrichment of cultivated soils in Southern Poland," Polish Journal of Environmental Studies, vol. 12, no. 2, pp. 187–192, 2003.

5.   P. J. Temple, S. N. Linzon, and B. L. Chai, "Contamination of vegetation and soil by arsenic emissions from secondary lead smelters," Environmental Pollution, vol. 12, no. 4, pp. 311–320, 1977.

6.   D. A. Bright, B. Coedy, W. T. Dushenko, and K. J. Reimer, "Arsenic transport in a watershed receiving gold mine effluent near Yellowknife, Northwest Territories, Canada," Science of the Total Environment, vol. 155, no. 3, pp. 237–252, 1994.

7.   L. E. Hunt and A. G. Howard, "Arsenic speciation and distribution in the Carnon estuary following the acute discharge of contaminated water from a disused mine," Marine Pollution Bulletin, vol. 28, no. 1, pp. 33–38, 1994.

8.   R. G. McLaren, R. Naidu, and K. G. Tiller, "Fractionation of arsenic in soils contaminated by cattle dip," in Proceedings of the 1st International Conference: Contaminants and the Soil Environment, p. 177, Adelaide, Australia, 1996.

9.   M. Azcue, A. Mudroch, F. Rosa, and G. E. M. Hall, "Effects of abandoned gold mine tailings on the arsenic concentrations in water and sediments of Jack of Clubs Lake, BC," Environmental Technology, vol. 15, no. 7, pp. 669–678, 1994.

10. C. M. Walsh and D. R. Keeney, Behavior and Phototoxicity of Inorganic Arsenic in Soils, vol. 7 of ACS Symposium Series, American Chemical Society, Washington, DC, USA, 1975.

11. B. K. Mandal and K. T. Suzuki, "Arsenic round the world: a review," Talanta, vol. 58, no. 1, pp. 201–235, 2002.

12. V. Straskraba and R. E. Moran, Environmental Occurrence and Impacts of Arsenic at Gold Mining Sites in the Western United States, Mine Water and the Environment, International Mine Water Association, 2006, http://www.imwa.info/.

13. R. S. Oremland and J. F. Stolz, "The ecology of arsenic," Science, vol. 300, no. 5621, pp. 939–944, 2003.

14. United States Environmental Protection Agency, Arsenic in Drinking Water, 2010, http://water.epa.gov/lawsregs/ rulesregs/ sdwa/arsenic/index.cfm.

15. Pollution Control Department-Ministry of Natural Resources and Environment, Notification of Thailand National Environmental Board No. 25, 2004.

16. American Society for Testing and Materials, Annual Book of ASTM Standards, vol. 4, West Conshoho, Pa, USA, 2000.

17. D. A. Skoog, F. J. Holler, and S. R. Crouch, Principle of Instrumental Analysis, Thomson Brooks/Cole, Belmont, Calif, USA, 6th edition, 2007.

18. R. M. Bailey, S. Stokes, and H. Bray, Inductively-Coupled Plasma Mass Spectrometry (ICP–MS) for Does Rate Determination: Some Guidelines for Sample Preparation and Analysis, Oxford Luminescence Research Group, School of Geography and the Environment and University of Oxford, Oxford, UK, 2003.

19. Department of Sustainable Natural Resources, Unified Soil Classification System, http://www.environment.nsw.gov.au/ resources/soils/testmethods/usc.pdf.

20. Mitchell, The ESRI Guide to GIS Analysis Volume 3; Modeling Suitability Moment and Interaction, ESRI, California, Calif, USA, 2012.

21. K. Krivoruchko, Spatial Statistical Data Analysis for GIS User, ESRI, Redlands, Calif, USA, 2011.

22. K. Johnston, ArcGIS 9: Using ArcGIS Geostatistical Analyst, Esri Press, 2003.

23. E. Holzbecher, Environmental Modeling Using MATLAB, Springer, Berlin, Germany, 2007.

Chapter 2

The Mechanism of Wellbore Weakening in Worn Casing-Cement-Formation System

Zheng Shen[1], Frederick E. Beck[1], and Kegang Ling[2]

[1]Texas A&M University, College Station, TX 77843, USA
[2]University of North Dakota, Grand Forks, ND 58202, USA

ABSTRACT

Maintaining casing integrity, in terms of downhole zonal isolations and well stability, is extremely important in oil/gas wells. Casing wear occurs not only in directional drilling, but also in vertical drilling with a slight deviation angle. In most hydrocarbon wells, deteriorated casing was reported from the onset of casing wear by the presence of friction force during the rotation of drillpipe.

The friction force against the casing wall causes the reduction of casing strength. Furthermore, the rotation of drillpipe combined with corrosive drilling fluids could dramatically degrade the casing strength. We used a finite element analysis to focus on the stress evolution in worn casings. Comparison study between worn casing and perfect casing was conducted. Our study showed that the thermal load significantly increases the stress concentration of the worn casing in the wellbore. Finite element solutions indicated that the radial stress of the worn casing is not affected as much as the hoop stress. Along with the increased burst pressure or the elevated temperature, the unworn portion of the casing also suffers from severe compression stress. This work is important to broadening the understanding of well engineers through addressing the true stress profile of worn casing in cemented wellbore.

INTRODUCTION

Casing wear in the oil and gas industry is recorded on a world basis. Rotation of drillpipe during the drilling process creates significant contact forces that result in the reduction of casing wall thickness. Thickness reduction of the casing wall weakens the burst and collapse resistances of casing, where stress concentration at the worn location is expected. Casing wear will be accelerated in the presence of corrosive fluids.

Casing wear is a serious problem that is not limited to directional or extended wells. Because the contact pressure generated on the inner surface of the casing becomes much harder to control during the drill bit penetration into deep formations, worn casing is also found in vertical wells. The strength of the casing depends on the material and geometry of the casing; the calculation of the burst and collapse pressures of casing is given in the Appendix. All criteria indicate that the burst and collapse resistances strengthen with the increase of casing wall thickness.

Within the limited information about the stress profile of worn casings, casing wear is analyzed alone in models without including cement sheath and formation. The stress profiles of worn casing developed by using the existing isothermal models [1] are insufficient to provide the whole picture regarding how worn casing behaves in a downhole field condition. The drilling process becomes more complicated in deep formations where the temperature of wellbore fluids is significantly different in conventional wells. In event of the worn casing, maintaining the stability of the cased wellbore in high-pressure and high-temperature (HPHT) wells is challenging because of large potential variations of temperature and pressure. High temperature can bring significant pressure increase in the sealed annuli, which can lead to further failures of the casing string or the production liner [2].

Casing wear can be caused by different parts of drilling tools. Casing wear typically results from contact pressure of the casing and the tool joint. As shown in Figure 1, the tool joint represented by part 2 has a relatively small diameter compared to the casing represented by part 1. The thickness of casing wall decreases because of the crescent-shaped wear by the tool joint indicted in the red area. After the reduction of casing wall thickness, stress caused by wellbore fluid pressure and formation in situ stresses is expected to be concentrated on the worn casing.

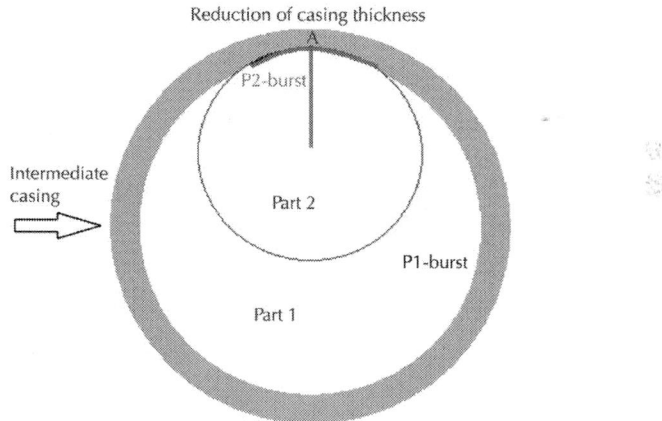

Figure 1: Crescent shape worn casing by the tool joint.

Thorough understanding of the true stress profile of worn casing leads to economic and safe casing design. Unfortunately, mechanical behaviors of casing from past studies were incomplete because of ignoring the formation and time-dependent temperature effect.

Researchers have investigated the stress profile of worn casing, focusing on the standalone casing [3, 4]. Others found the relation of wear depth and contact force through experimental tests, with each wear depth recorded at a given contact force [5]. Researchers [6, 7] have also studied the effect of formation to casing. They concluded that the surrounding formation significantly increased the burst resistance of casing, but the work was applied to the isothermal downhole condition. None of the existing studies considered the thermal and mechanical behaviors of worn casing in a cemented well. The mechanism of wellbore weakening due to worn casing has not yet been sufficiently understood. The limitations of the existing models motivated us to determine true stress profile in and around worn casing in the cased hole.

This work aims to describe the stress concentration of casing after wear in a cemented well. The results showed that the formation in situ stresses and temperature largely impact the stress profile of the worn casing in the cemented wellbore. Casing uncemented with formation suffers from tension when the burst pressure is applied in the wellbore. Without considering the temperature effect in the worn casing, the stress concentration of the worn part is apparently underestimated. The traditional Lame equations for stress calculation in the wellbore are weakened by not including the temperature effect, cement sheath, and formation. The results also showed that the elevated temperature has expanded the worn casing and resulted in further compression in the worn casing. This work established the mechanical responses of worn casing through addressing the true stress profile of worn casing in cemented wellbore. The findings of the present work will be beneficial to well engineers.

PREVIOUS STUDIES

Uncemented casing suffers from serious tension when the burst pressure is applied on the inner wall of casing. For a perfect casing without wear in an isothermal condition, the radial stress and hoop stress can be evaluated using (1). The equations do not take into account the effects of cement sheath and formation:

$$\sigma_r = \frac{p_w r_w^2}{r_o^2 - r_w^2}\left(1 - \frac{r_o^2}{r^2}\right) - \frac{p_o r_o^2}{r_o^2 - r_w^2}\left(1 - \frac{r_w^2}{r^2}\right),$$

$$\sigma_\theta = \frac{p_w r_w^2}{r_o^2 - r_w^2}\left(1 + \frac{r_o^2}{r^2}\right) - \frac{p_o r_o^2}{r_o^2 - r_w^2}\left(1 + \frac{r_w^2}{r^2}\right).$$

$$(1)$$

The reduction of casing wall thickness mainly results from casing corrosion and casing wear. It is well known that drilling through sand formation zones causes serious erosion randomly along the casing pipe. Casing corrosion is categorized into several types. Decarburization is a hydrogen-carbon attack of a component of any alloy under high temperature. Galvanic corrosion occurs when two dissimilar metals are in contact. This type of corrosion, most likely occurring in the surface casing, is normally seen as a result of the active macrocorrosion cells when the surface casing is not well cemented. Biological corrosion happens as a result of activities of living organisms. This happens in the environment of temperature ranging from 30°F to 180°F, pH measurement of 0 to 11, and pressures up to 15,000 psi [8].

In drilling, the rotation of drillpipes could uniformly reduce the casing thickness from the inner surface of casing or create a crescent-shaped wear pattern. Researchers have focused on the wear volume of casing through laboratory measurements [9]. Casing wear can be caused by tool joints, drillpipe, and wireline. Casing wear by the drillpipes and wirelines is not as significant as that by the tool joints. Casing wear by tool joints is determined by the factors such as drill string rotation time and the speed, drilling mud properties, casing strength, and well dogleg severity.

Stress concentration around worn casing has been studied by some researchers. A casing wear model is largely weakened by not considering the temperature and formation. Researchers [1] focused on the rupture capacity of casing after wear. The analytical solution for the hoop stress at the surfaces of worn casing was constructed by dividing the entire worn casing into three superimposable shapes. To obtain the induced hoop stress of the worn casing, the superposition principle was adopted in a bipolar coordinate. Others argued that casing wear models can be simplified by assuming a slotted ring in the casing inner wall [3]. In their method, the resistance of the hoop stress is decreased because of the reduction of casing wall plus the extra burst pressure acting on the surfaces of the slotted ring. The stress under the worn surface was also investigated. The experimental work indicates that the area below the wear surface experiences unusual strain [10].

Researchers investigated the maximum wear groove depth resulting from the contact pressure applied to the inner wall of casing [4]. Various-size drill strings were used to find a relation for the groove depth of crescent-shaped casing wear as a function of time. The casing wear was tested in single, sharp, and blunt groove forms. Another study emphasized the alleviation and prevention of casing wear by using optimized well operations, such as dogleg severity control and sealant application in the worn sections [11].

NUMERICAL MODELING

Casing wear causes significant reduction of the casing strength. Worn casing is under the risk of tangential collapse and radial cracking in the wellbore system. Force equilibrium can be expressed in terms of the normal force, shear force, and body force. The general relation of the forces in a cylindrical coordinate system is given in (2) [12]. Ideally, an analytical solution for the hoop stress in the worn casing is required. Typical Lame equations used in casing integrity analysis do not qualify for the effects of worn shape and formation. It is difficult to develop a reasonable and accurate analytical model investigating worn casing. Because the existing force balance equations are not suitable for addressing the effects of cement sheath, formation, and temperature in worn casing system, it is difficult to develop a reasonable and accurate analytical model investigating worn casing:

$$\frac{\partial \sigma_r}{\partial r} + \frac{1}{r}\frac{\partial \tau_{r\theta}}{\partial \theta} + \frac{\sigma_r - \sigma_\theta}{r} + \sigma f_r = 0.$$

(2)

Alternatively, a finite element model can be used to overcome the limitations of the analytical models. Finite element analysis (FEA) is a computer-based numerical technique applied to many engineering problems. This method is popular in performing prejob designs for primary jobs or postmodified jobs for remedial actions. By using a finite element method, a structure, such as the casing-cement-formation system, can be broken down into small elements composed of nodes in a global coordinate system. The deflections and stresses of all nodes must be calculated using computer programs because of the complexity of the global system. The computation time of solving for the finite element model depends on the number of total elements and computer performance.

The numerical simulator ANSYS was used. Numerical simulations were performed to analyze thermal and mechanical behaviors of worn casing in the cemented wellbore. In the present model, the worn casing was located in a vertical well and was perfectly cemented with the surrounding formation. Plane strain assumes that one direction of the investigated body is much larger than the other two directions. In our case, the vertical section of the wellbore is much larger than the radial direction. It is reasonable to construct a 2D numerical model based on the concept of plane strain, which leads to the following relations in (3). The strain of normal to the x-y plane and the shear strains γ_{xz} and γ_{yz} are zero:

$$\varepsilon_z = \gamma_{xz} = \gamma_{yz} = 0.$$

(3)

Drilling fluids have significant effect on the stability of the wellbore system. Because of the geometry of the casing cement formation, the thermal energy exchange in the near-wellbore system reaches equilibrium in a short time. A past study [13] has shown that wellbore fluids induce major stress change to the casing in the wellbore at the first hour. In this model, the minimum horizontal stress is equal to maximum stress. Fluids are pumping into the wellbore and circulating back to the surface through the wellbore annulus. The duration of the circulation is one hour. Wellbore fluid temperature is different from the formation temperature. The pressure and temperature at the boundaries are specified in the following:

$$r = r_w, \qquad \sigma = p_w,$$

$$r = r_o, \qquad \sigma = p_o,$$

$$r = r_w, \qquad T = T_w,$$

$$r = r_o, \qquad T = T_o. \tag{4}$$

Accuracy of a finite element scheme measures the closeness between the analytical solution and the numerical solution. We first simplified the boundary conditions of the numerical model we used; thus the analytical solutions to the same problem can be found to verify the numerical results. Figures 2 and 3 show the radial stress and tangential stress of casing subject to the internal burst pressure of 1000 psi while ignoring the formation effect. The numerical solution demonstrated a good match compared with the analytical result.

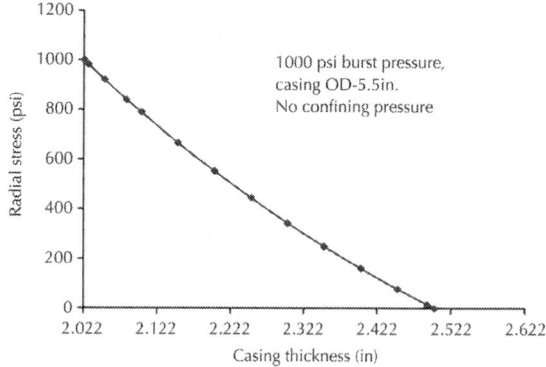

Figure 2: Casing radial stress in the radial direction.

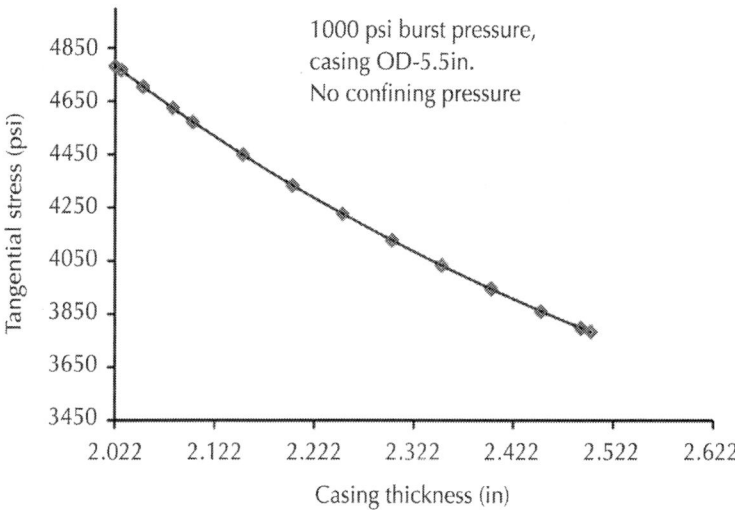

Figure 3: Casing tangential stress in the radial direction.

In this following study, the cased wellbore diameter of 6 in. is used, and the outer diameter of the formation is 10 times as large as the wellbore diameter. Cohesion force and friction angle of cement and rock are included in the simulations. The cohesion force is the force of attraction between the molecules of the same substance, and the friction angle measures the ability of rock or cement to resist the shear stresses. Other material properties of the worn casing cement formation used in the numerical model are listed in Tables 1 and 2.

Table 1: Geometric and geomechanical properties

	ID (in.)	OD (in.)	Elastic modulus (psi)	Poisson ratio	Cohesion force (psi)	Friction angle (°)
Casing	6	7	2.80E + 07	0.3	Not applicable	Not applicable
Cement sheath	7	9	3.00E + 06	0.24	8.00E + 03	30
Formation	9	60	6.00E + 06	0.2	6.00E + 03	30

Table 2: Thermal properties

	Density (lb/in³)	Conductivity (btu/in-hr-F)	Heat capacity (Btu/lb-F)	Thermal expansion (F⁻¹)
Casing	0.253	0.6	0.12	1.8E − 05
Cement sheath	0.065	0.05	0.5	1.2E − 05
Formation	0.072	0.04	0.4	1.4E − 05

A crescent shape of worn casing is assumed as shown in Figure 3. The reduction factor of worn casing is defined as the biggest reduction of casing thickness over the original casing thickness in (5). Field observations have shown that worn casing in many wells lost approximately 20% of the casing thickness [1]. The original casing thickness is 0.5 in. In the deepest worn point, 0.1 in. thickness of casing is lost after casing wear:

Table 3 lists the fluid temperatures and pressures under different scenarios. To evaluate the stress concentration in the worn casing, six base models were built with either different temperatures or wellbore pressures in the model. Case 1 simulates the perfect casing-cement-formation system without casing wear. Case 2 through 6 are used for the worn casing-cement-formation system in different wellbore fluid pressure and temperature. The results analysis is presented in the next section.

Table 3: Fluid temperature and burst pressure for six cases

	Wellbore fluid temperature (°F)	Formation temperature (°F)	Initial temperature (°F)	Burst pressure (psi)	Rock in situ stress (psi)	Casing
Case1	350	350	350	12,500	11,000	Perfect
Case2	350	350	350	12,500	11,000	Worn
Case3	450	350	350	12,500	11,000	Worn
Case4	450	350	350	13,500	11,000	Worn
Case5	450	350	350	14,500	11,000	Worn
Case6	450	350	350	13,500	0	Worn

RESULTS

The results presented are near-wellbore mechanical behaviors when the wellbore fluids flow in the worn casing-cement-formation system. Although stresses and displacements at any location can be found using this model, our discussions focus on the casing-cement case. Contour figures are used to show the stress distributions in different downhole conditions. In the model, a negative sign represents the compression and a positive sign is for tension.

Case 1: The aim of this case is to validate this model. The results presented in Figures 4 and 5are consistent with the existing solutions. In previous studies, different researchers predict the maximum radial stress at the casing-cement interface in a perfect casing-cement system [6]. In Case 1, the perfect casing without thermal effects and wear is analyzed. The figures reveal the contour profiles of radial stress and hoop stress in the casing-cement part. The maximum radial stress reaches approximately 13,600 psi, which is uniformly distributed at the casing-cement interface. The inner wall of the casing suffers from the highest hoop stress.

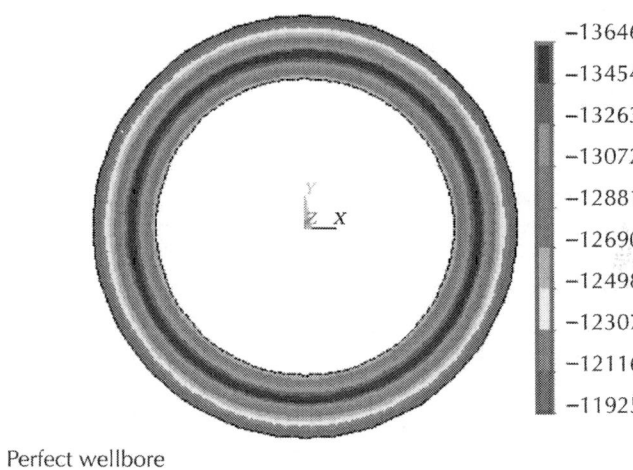

Perfect wellbore

Figure 4: Radial stress inside the perfect casing cement for Case 1.

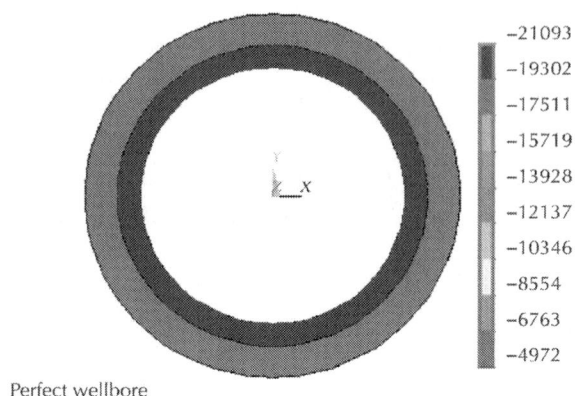

Perfect wellbore

Figure 5: Hoop stress inside the perfect casing cement for Case 1.

Case 2: The reduction of casing thickness decreases the casing strength. No heat transfer occurs in Case 2 because the temperatures of the fluid and rock are identical. Although the casing is worn, the maximum radial stress still occurs at the interface of casing and cement. There is little difference between the radial stresses in the worn part and those in the unworn part, as shown in Figure 6.

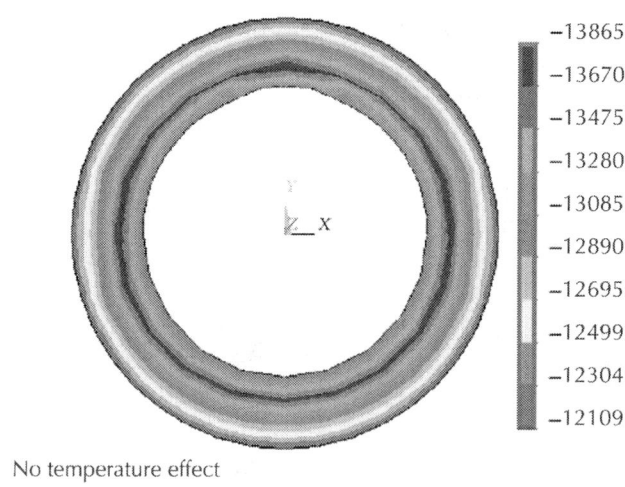

No temperature effect

Figure 6: Radial stress inside the worn casing cement for Case 2.

The tension failure of worn casing will be overestimated without considering the confining effect of cement and rock. A previous study shows that the wellbore burst pressure causes severe tangential tension at the worn casing [3]. This occurs because the burst pressure in the uncemented wellbore produces the tension only when the casing stands alone. Figure 7 shows that the large compression hoop stress occurring around the worn area likely causes the yield of worn casing.

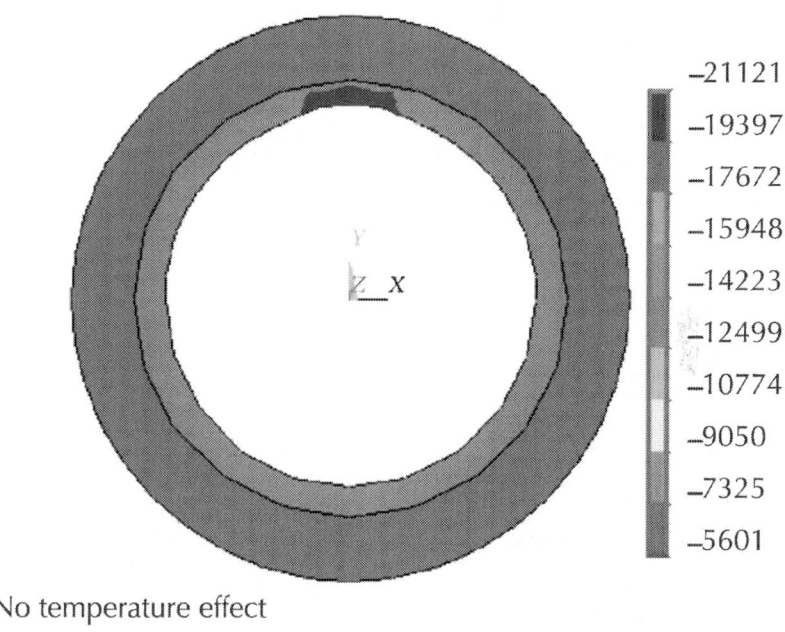

–21121
–19397
–17672
–15948
–14223
–12499
–10774
–9050
–7325
–5601

No temperature effect

Figure 7: Hoop stress inside the worn casing cement for Case 2.

Case 3: The fluid temperature is 450°F, which is much higher than the rock temperature of 350°F. Thermal stress acts on the casing-cement part until the heat transfer reaches equilibrium. Compared with the maximum radial stress of 13,866 psi in Case 2, maximum radial stress in the casing-cement part is increased by one-third because of the thermal stress (Figure 8).

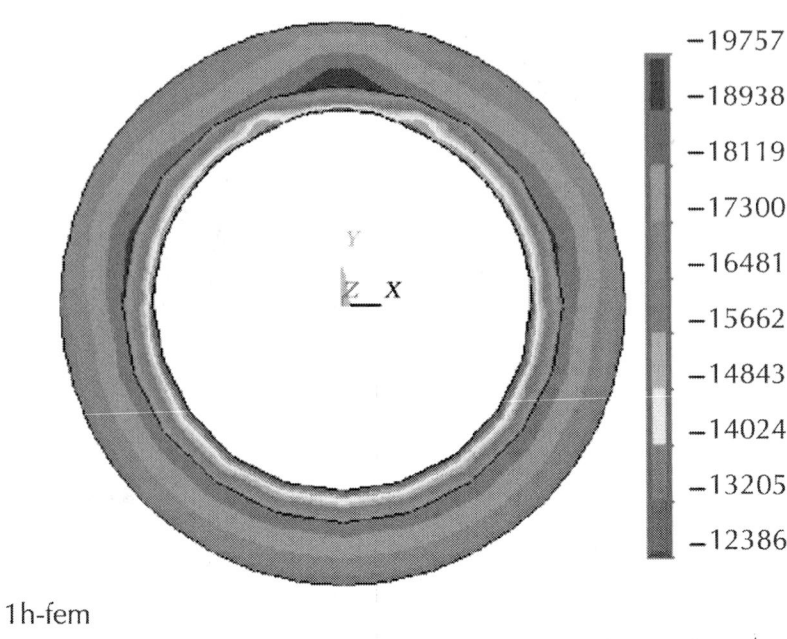

1h-fem

Figure 8: Radial stress inside the worn casing cement for Case 3.

The casing is still constrained by the cement and formation, but the elevated temperature in the casing-cement-formation system resulting from the wellbore fluids tends to cause expansion of the worn casing. The results shown in Figures 8 and 9 indicate that the consequence of the casing expansion is to produce further compression on the casing-cement part. The maximum hoop stress of 73,791 psi in Case 3 is close to three times larger than the 21,121 psi in Case 2. It is concluded that the increase of hoop stress is much more than the increase of radial stress in the worn casing cement due to the thermal effect. Several types of casings will fail under this circumstance. The minimum yield strength of casing strings such as J55 and K55 is 55,000 psi. The high temperature difference in the wellbore could put the worn casing-cement-formation system under high risk of failure.

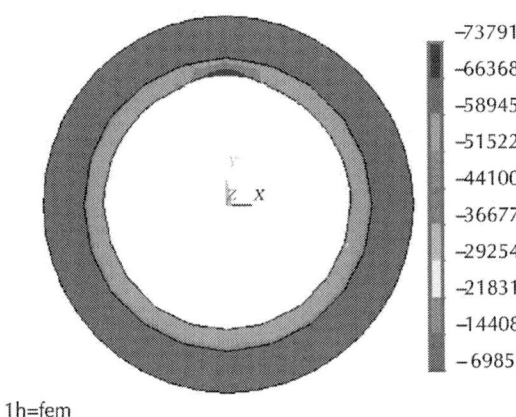

1h=fem

Figure 9: Hoop stress inside the worn casing cement for Case 3.

Case 4: The temperature profile is unchanged. The burst pressure of 13,500 psi is used. The compression radial stress in the worn casing cement is slightly larger than that in Case 3, as shown in Figure 10. This indicates that the larger the wellbore burst pressure, the greater the induced casing radial stress resulting from the higher fluid pressure acting on the inner wall of the wellbore.

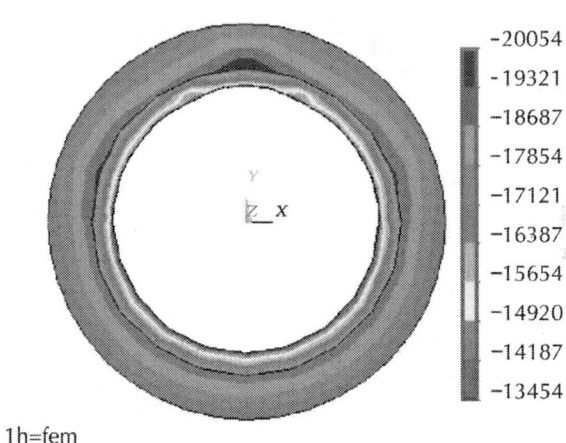

1h=fem

Figure 10: Radial stress inside the worn casing cement for Case 4.

It is interesting to note that the hoop stress inside the worn casing cement caused by the burst pressure of 13,500 psi is lower than that by the burst pressure of 12,500 psi, as shown in Figure 11. This hoop stress is decreased from 73,791 psi to 68,671 psi. The thermal expansion of casing in the constrained environment due to the cement and formation causes extremely large compression tangential stress, whereas the increased burst pressure tends to expand the worn casing in the opposite direction. It is concluded that a higher burst pressure in the wellbore could reduce the maximum hoop stress in the worn casing-cement part.

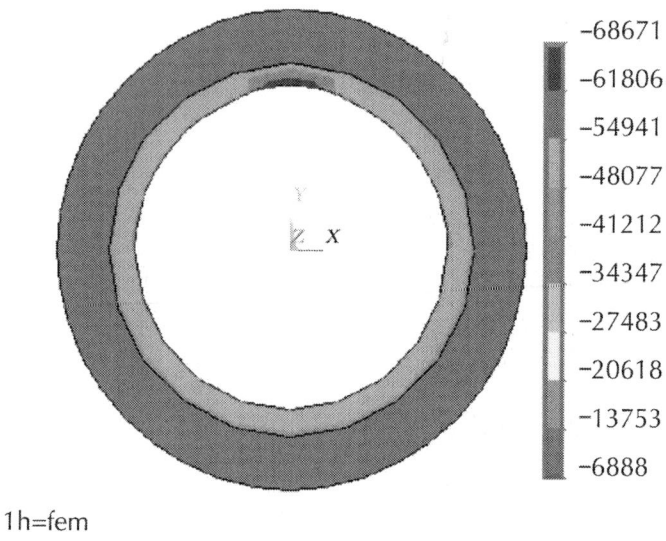

1h=fem

Figure 11: Hoop stress inside the worn casing cement for Case 4.

Case 5: Figures 12 and 13 reveal the radial stress and hoop stress profile in the worn casing-cement system. The temperature profile in Case 5 is identical with that used in Cases 2 and 3. The burst pressure increases to 14,500 psi. The results show that the radial stress in the system slightly increases with a bigger burst pressure. As expected, the maximum hoop stress occurs on the worn part, and it decreases by 5,000 psi along with the increase of burst pressure.

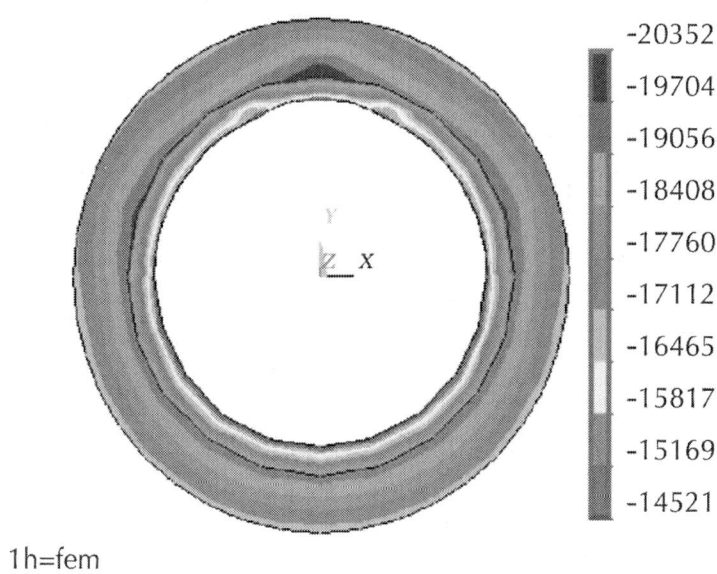

1h=fem

Figure 12: Radial stress inside the worn casing cement for Case 5.

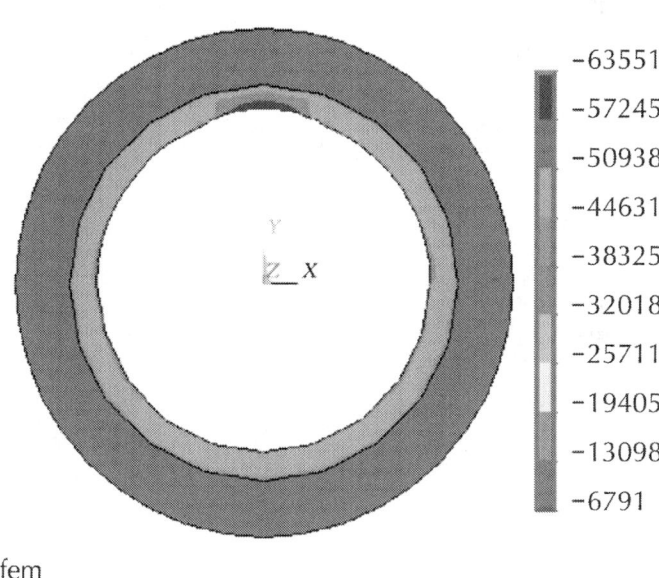

1h-fem

Figure 13: Hoop stress inside the worn casing cement for Case 5.

Cases 2 to 5 present stress profiles of the worn casing under different burst pressures. The stress concentration on the worn part in the casing-cement-formation is predicted in all cases. The risk of tangential compression failure in the worn casing decreases along with the increase of burst pressure.

Case 6: Figures 14 and 15 show the radial stress and hoop stress of the worn casing cement, respectively. The results reveal that the formation is important to prevent tension failure on the worn casing. To better illustrate mechanical behaviors of the worn casing caused by the cement and formation, zero formation in situ stress is used in this case. The maximum radial stress occurs on the inner wall of the worn casing, which differs from Cases 1 to 4. The maximum hoop stress, as shown in Figure 15, occurs on the worn part. The worn casing suffers from extreme tension without the formation in situ stress.

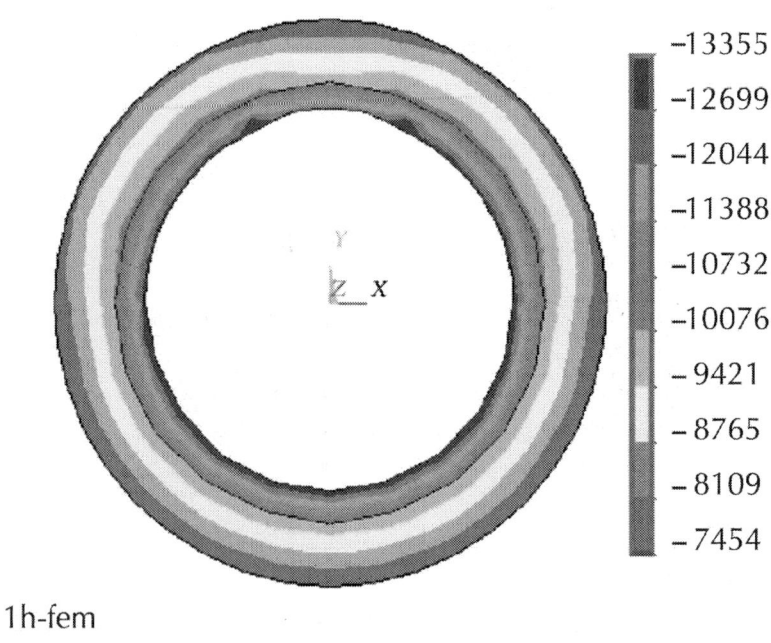

1h-fem

Figure 14: Radial stress inside the worn casing cement for Case 6.

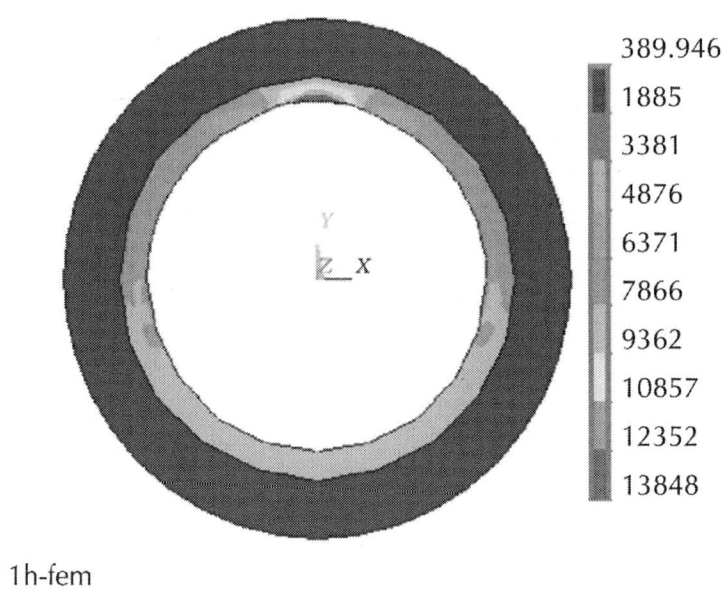

	389.946
	1885
	3381
	4876
	6371
	7866
	9362
	10857
	12352
	13848

1h-fem

Figure 15: Hoop stress inside the worn casing cement for Case 6.

CONCLUSIONS

The present numerical model has addressed the complete wellbore system consisting of wellbore fluid, casing, cement, and formation. Two principal stresses are analyzed in different downhole conditions. A worn casing-cement-formation model is built using finite element analysis. Ignoring the effect of temperature, the induced stresses in the worn casing-cement-formation system were slightly higher than that in a perfect casing-cement-formation system. In this study, casing expansion due to the temperature difference between wellbore and formation produces severe compression on the worn casing-cement part. The results show that the induced stress could exceed the minimum yield strength of casing strings such as J55 and K55.

Although the casing thickness is partially reduced after wear, worn casing in the downhole system suffers from apparent stress concentration. It is concluded that the increase of the hoop stress is much more than the increase of radial stress in the worn casing cement due to the thermal effect. The burst pressure tends to cause the tension of worn casing. The effects of formation and temperature dominate the stress distribution of the worn casing. It is important to note that high wellbore burst pressures facilitate lowering the compression tangential failure in the worn casing-cement part in the cemented wellbore.

REFERENCES

1.   J. S. Song, J. Bowen, and F. Klementich, "Internal pressure capacity of crescent-shaped wear casing," in Proceedings of the IADC/SPE Drilling Conference, pp. 547–553, New Orleans, La, USA, February 1992.

2.   P. Oudeman and M. Kerem, "Transient behavior of annular pressure build-up in HP/HT wells," in Proceedings of the 11th ADIPEC: Abu Dhabi International Petroleum Exhibition and Conference, pp. 665–674, Abu Dhabi, UAE, October 2004.

3.   J. Wu and M. G. Zhang, "Casing burst strength after casing wear," in Proceedings of the SPE Production and Operations Symposium 2005, pp. 517–526, Oklahoma City, Okla, USA, April 2005.

4.   R. W. Hall Jr. and K. P. Malloy Sr., "Contact pressure threshold: an important new aspect of casing wear," in SPE Production and Operations Symposium 2005: Anticipate the Future, Build on the Present, Celebrate the Past, pp. 499–505, Society of Petroleum Engineers, Oklahoma City, Okla, USA, April 2005.

5.   D. Gao, L. Sun, and J. Lian, "Prediction of casing wear in extended-reach drilling," Petroleum Science, vol. 7, no. 4, pp. 494–501, 2010.

6.   W. W. Fleckenstein, A. W. Eustes III, and M. G. Miller, "Burstinduced stresses in cemented wellbores," SPE Drilling and Completion, vol. 16, no. 2, pp. 74–82, 2001.

7.   W. J. Rodriguez, W. W. Fleckenstein, and A. W. Eustes, "Simulation of collapse loads on cemented casing using finite element analysis," in Proceedings of the SPE Annual Technical Conference and Exhibition, pp. 5239–5247, Denver, Colo, USA, October 2003.

8.   S. Talabani, B. Atlas, M. B. Al-Khatiri, and M. R. Islam, "An alternate approach to downhole corrosion mitigation," Journal of Petroleum Science and Engineering, vol. 26, no. 1-4, pp. 41–48, 2000.

9.   W. B. Bradley and J. E. Fontenot, "The prediction and control of casing wear (includes associated papers 6398 and 6399)," SPE Journal of Petroleum Technology, vol. 27, no. 2, pp. 233–245, 1975.

10.   W. M. Rainforth, "Microstructural evolution at the worn surface: a comparison of metals and ceramics," Wear, vol. 245, no. 1-2, pp. 162–177, 2000.

11.   B. Calhoun, S. Langdon, J. Wu, P. Hogan, and K. Rutledge, "Casing wear prediction and management in deepwater wells," in Proceedings of the SPE Deepwater Drilling and Completions Conference 2010, pp. 232–241, Society of Petroleum Engineers, Galveston, Tex, USA, October 2010.

12.   E. Fjar, R. M. Holt, A. M. Raaen, R. Risnes, and P. Horsrud, Petroleum Related Rock Mechanics, Elsevier, Oxford, UK, 2004.

13.   Z. Shen and F. E. Beck, "Three-dimensional modeling of casing and cement sheath behavior in layered, nonhomogeneous formations," in Proceedings of the IADC/SPE Asia Pacific Drilling Technology Conference 2012, pp. 949–958, Society of Petroleum Engineers, July 2012.

14.   T. Tamano, "A new empirical formula for collapse resistance of commercial casing," in ASME Transactions of Energy Resources Technology, 2005.

Chapter 3

Planning for Reliable Coal Quality Delivery Considering Geological Variability: A Case Study in Polish Lignite Mining

Wojciech Naworyta[1], Szymon Sypniowski[2], and Jörg Benndorf[3]

[1]Department of Surface Mining, AGH University of Science and Technology, Mickiewicza Avenue 30, 30-059 Krakow, Poland

[2]Department of Mineral Resources Acquisition, MEERI PAS, Wybickiego Street 7, 31-261 Krakow, Poland

[3]Faculty of Civil Engineering and Geoscience, Delft University of Technology, Building 23, Stevinweg 1, 2600 GA Delft, Netherlands

ABSTRACT

The aim of coal quality control in coal mines is to supply power plants daily with extracted raw material within certain coal quality constraints. On the example of a selected part of a lignite deposit, the problem of quality control for the run-of-mine lignite stream is discussed. The main goal is to understand potential fluctuations and deviations from production targets dependent on design options before an investment is done. A single quality parameter of the deposit is selected for this analysis—the calorific value of raw lignite. The approach requires an integrated analysis of deposit inherent variability, the extraction sequence, and the blending option during material transportation. Based on drill-hole data models capturing of spatial variability of the attribute of consideration are generated. An analysis based on two modelling approaches, Kriging and sequential Gaussian simulation, reveals advantages and disadvantages lead to conclusions about their suitability for the control of raw material quality. In a second step, based on a production schedule, the variability of the calorific value in the lignite stream has been analysed. In a third step the effect of different design options, multiple excavators and a blending bed, was investigated.

INTRODUCTION

Environmental and economic considerations in the electrical energy industry rise the necessity to constantly improve the efficiency of power units. One way to increase the efficiency of energy production in the power plants based on fossil fuels is to supply the raw material with specific and relatively stable quality parameters.

In the case of lignite, the spatial variability of parameters is quite large. Given the variability criterion, lignite belongs to the second group of deposits in the Polish classification. The coefficient of variation v [%] is defined as the ratio of the standard deviation to the mean value of the basic parameters and is usually in the range of 30% to 60%. The exception is the calorific value which has a relatively low volatility in the range of 9 to 16% [1].

To meet customer's requirements, the planning and design of a mining operation have to focus on technical and operational measures to reduce the in situ variability of critical coal attributes during mining and material handling. The aim of different design options, such as the use of blending beds or multiple excavators simultaneously, is to transform the in situ variability in the deposit to a level which meets customers' requirements. For investigating the effect of a coal blending beds the theory of variance reduction in bed blending is well established (e.g., [2]). It is based on the variogram transformation of the incoming to the outgoing stream. Several documented applications (e.g., [3, 4]) use techniques of stochastic simulation based on variograms of critical elements to simulate the variability of incoming material flows and to optimise the transformation process. Considering geologically more complex deposits this approach may be too simplified. To investigate the homogenisation effects in a continuous mining system, the deposit characteristics, in particular the local variability has to be linked with the extraction method, the mining sequence, and blending options throughout the operation [5].

In order to maintain stable raw material parameters, certain measures are undertaken referred to as the lignite stream quality management (e.g., [5–9]). This process begins with the exploration and documentation of the deposit and is conducted until the end of the mine's life. Coal quality control can be divided into several stages:

- identification of critical parameters and modelling of the deposit:
    a. identification and analysis of critical coal quality parameter,
    b. spatial modelling of the variability of quality parameters,
- mine planning (long-term planning):

    a. determination of the location ultimate pit limit and of the opening box cut,

    b. design of blending options and facilities, such as stock and blending yards,

    c. establishment of a long-term mining sequence and advances of the mining faces in time,

- exploitation and production control (operational planning):

    a. short-term production scheduling for the extraction equipment,

    b. prediction and online analysis of the quality of the extracted coal,

    c. logistics and transportation,

    d. storage and homogenisation of the raw material.

The analysis presented here relates to the second and third stage of the control process—the design of blending options and operational planning. The following sections will first investigate different geostatistical modelling approaches for their suitability to map realistically spatial variability of lignite attribute considered. In the second part the variability of the extracted material flow is evaluated, including bed-blending and multiple excavators, leading to design options for improved coal quality management and a reliable supply of the power plant.

This paper is a continuation of the aspects related to coal quality management in lignite mines discussed by the authors in previous publications. In particular, methods of conditional simulation in geostatistics investigated in [10, 11] are applied to full scale reserve modelling of a large lignite field in Poland aiming to understand variability of coal quality attributes at a short-term scale. Using these models the second part focuses on design issues of a stock and blending bed to understand its ability to control short-term variation. Contrary to the work described in [5, 12], which focuses on operational optimization of a coal stock and blending bed, here the aim is to understand the effect of the bed size to control coal quality fluctuations of final products to be sold. The combined approach discussed in this paper allows decisions on the optimal stock and blending bed design to be evaluated in the design phase, before short-term operation is actually executed and real fluctuations experienced.

THE OBJECTIVE OF THIS CASE STUDY

For the process of lignite quality control at the stage of operational planning it is necessary to have sufficient exploration information about the deposit. In the mines this task is accomplished in different ways. One of them is to explore the deposit with drill-holes drilled from the roof of the exposed lignite—the so called operational exploration. The holes in the deposit analysed in this paper were drilled in a dense grid of 50 by 50 meters. Although, in comparison to the geological documentation stage, the operational exploration has a higher information content, the actual parameters of the mined lignite still often differ from values identified during this drilling period.

The main objective of this study is to understand possible deviations with respect to the expected calorific of coal produced based on operational exploration data, that is, to assess to what extent these data provide accurate information for the tasks related to the quality control of the mined mineral. To achieve this goal and test the suitability of different approaches, two methods of geostatistical modelling are compared, ordinary Kriging and conditional simulation (e.g., [1]).

In a second step two different design options are investigated focusing on the effect of variability of run-of-mine lignite, which are as follows:

- the availability of a coal stock and blending yard for bed blending: different sizes are investigated,
- the availability of a second excavator and the possibility to blend two lignite streams on the belt conveyor.

For run-of-mine lignite quality control in the context of power plant supply multiple parameters such as calorific value, sulphur content, and silica content have to be taken into account. Without loss of generality, this paper focused on the analysis of the calorific value of the raw lignite Q_i^r.

THE METHOD

On the basis of the operational exploration within the area limits of six-month progress of extraction, variability models of the calorific value in particular mining blocks were created. This analysis was performed for the part of the deposit where the operational exploration is characterized by high regularity. Figure 1 shows the selected part of the deposit with respect to the entire deposit and the assumed mining progress in relation to all exploratory holes.

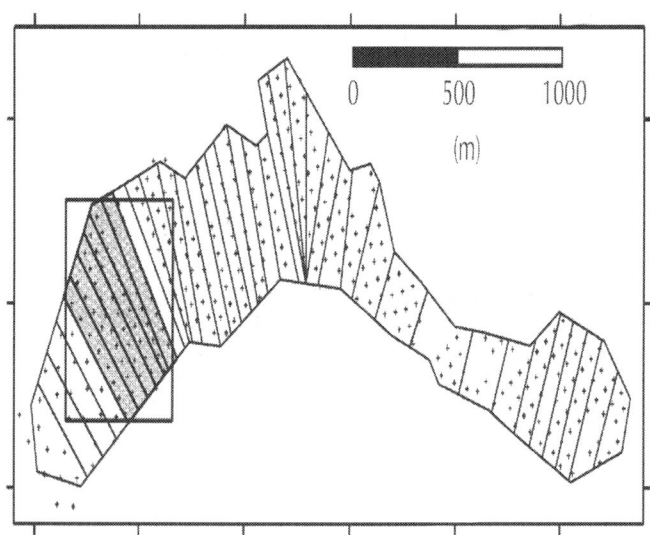

Figure 1: Location of the drill holes of the operational exploration and the limits of monthly mining progress. The rectangle marks the area selected for analysis.

On the basis of the calorific value variability models with a given mining direction, the variations of calorific value were calculated for a six-month period. Figure 2 shows a sequence of mining 195 consecutive exploitation blocks. Each mining block with dimensions of 30 × 30 meters corresponds to an actual average daily production of lignite from the analyzed deposit. With the average lignite seam thickness of approximately 6 meters and a density of 1.15 t/m³, a single exploitation block contains about 6,5 thousand tonnes of lignite.

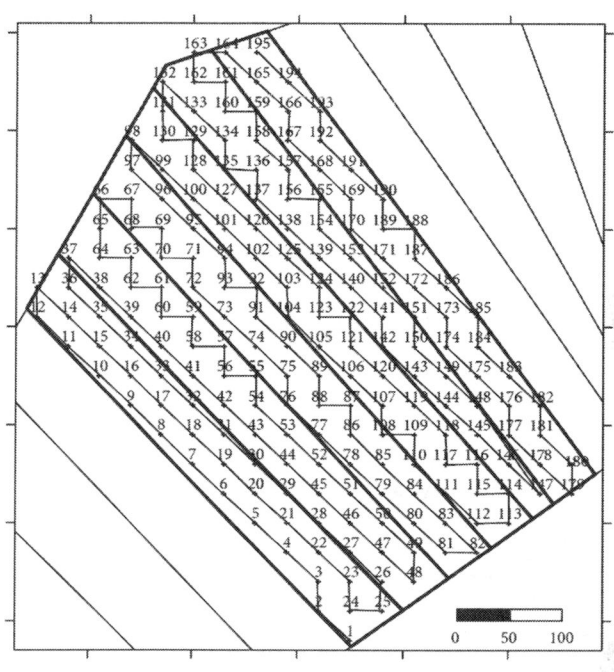

Figure 2: The order of exploitation within a six-month period. Each exploitation block has been marked with the number.

Based on the data the models of spatial variability of the calorific value in the deposit were created using the ordinary Kriging method and the direct sequential simulation method, which are implemented in the software S-GeMS [13]. The geostatistical simulation procedure is based on the idea of Monte-

Carlo simulation. Based on available observations of the deposit and on random numbers, the simulation can generate any number of models (herein referred to as realizations). The realizations are unique and at the same time characterized by identical probability to represent the actual deposit. All realizations accurately reflect the values at the observation points. Unlike ordinary Kriging, realizations resulting from simulation accurately reflect the statistical and structural features of the modelled parameters such as the density distribution and spatial variability. Local differences between particular realizations present the measure of uncertainty of the prediction conducted by the simulation on the basis of the available observations. In the paper 50 independent realizations of the calorific value for the selected part of the deposit are presented. Figure 4 shows two exemplifying realizations.

Both of the used methods require a variogram model capturing the spatial variability as input. First an empirical variogram is calculated and secondly a model is fitted. In the presented case the spherical basic structure resulted in the best fit (Figure 3, Table 1). Due to the lack of a clear directional variability in the modelled deposit, an omnidirectional variogram model was used.

Table 1: Basic features of variogram model of calorific value

Variogram model	Dimension and unit
Nugget effect	160 000 $(kJ/kg)^2$
Spherical model	235 000 $(kJ/kg)^2$
Autocorrelation range	900 m

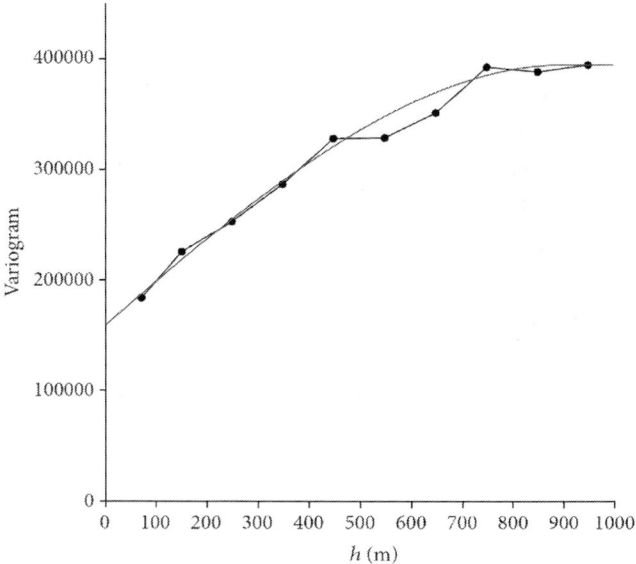

Figure 3: Experimental variogram with variogram model.

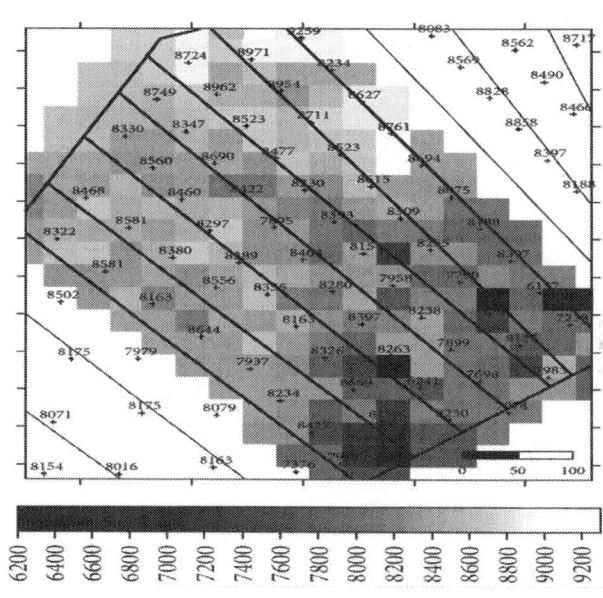

(a)

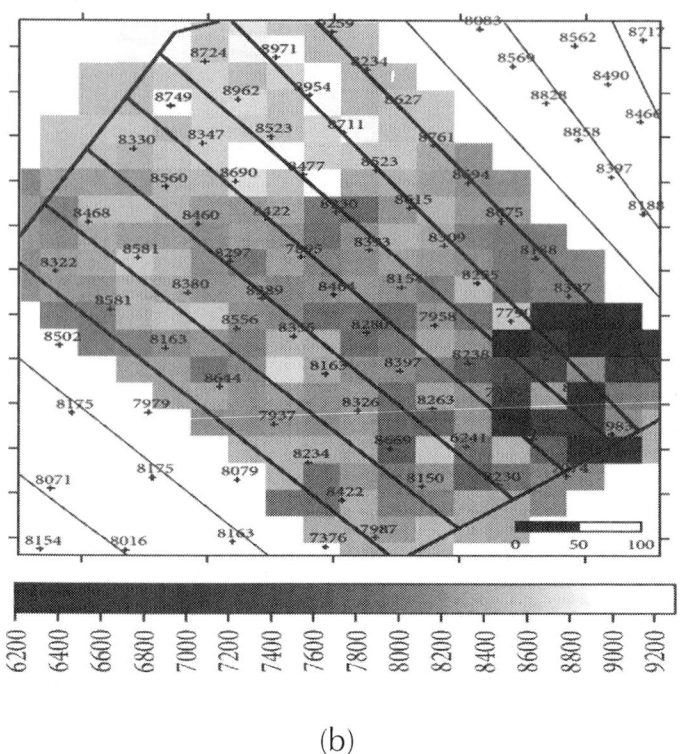

(b)

Figure 4: Models of calorific value Q_i^r—exemplifying realizations of geostatistical simulation.

THE DATA BASE USED FOR THE ANALYSIS

The variability models of calorific value were created based on 68 operational exploration drill holes located within the borders of mining and on the basis of the adjacent holes. Table 2 summarizes the basic statistical characteristics of the measurement data from 68 holes. As can be seen, both models perform well in reproducing the mean value of the drill holes. The variance of modelled blocks cannot directly be compared to the variance of exploration data, since both are based on a different support. However, it can be noticed that simulated block values appear more variable as Kriged block values. This effect results from the smoothing effect of Kriging.

Table 2: Basic statistics of calorific value based on 68 boreholes of the operational exploration and of the both models

	Data from the exploratory holes	Model ordinary Kriging	Model-exemplifying simulation
Number of holes/ number of blocks	68	195 (30 × 30 m)	195 (30 × 30 m)
The mean value	8267 kJ/kg	8274 kJ/kg	8260 kJ/kg
The standard devia-tion	521 kJ/kg	316 kJ/kg	485 kJ/kg
Coefficient of varia-tion	6,30%	3,80%	5,9%
The minimum value	6137 kJ/kg	7517 kJ/kg	6225 kJ/kg
The maximum value	9259 kJ/kg	8759 kJ/kg	9006 kJ/kg

RESULTS AND DISCUSSION

Figures 4 and 5 show the calorific value volatility models in the selected part of the deposit. To facilitate the assessment of the validity of the models, the figures also present the location of the operational exploration holes with their identified calorific value. Figure 4 shows two examples out of the total 50 conducted realizations of the simulation. The models differ from each other, and the differences are primarily in the blocks where there are no exploratory drill holes.

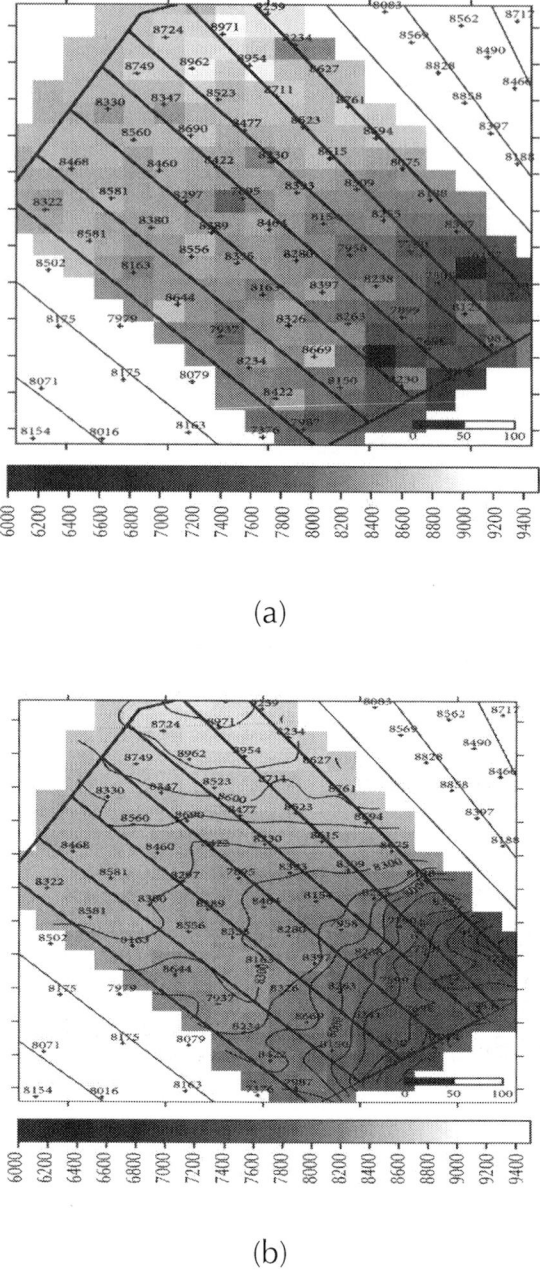

(a)

(b)

Figure 5: Variation of the calorific value Q_i^r —average of 50 realizations of the simulation (l) and ordinary Kriging (r).

Figure 5 summarizes the two models representing the expected spatial distribution of the calorific value—the average of the 50 realizations (a) and using the ordinary Kriging method (b). In the model created using the Kriging the calorific value changes gradually. The contour lines shown in Figure 5(b) show the effect of smoothing that occurred when using ordinary Kriging. Contrarily the variability is conserved in the single realization (Figure 4). When averaging all realizations, resulting in the so-called E-type estimator (Figure 5(a)), a very similar model to the one of Kriging is obtained.

Figure 6 presents the histograms of the calorific value models variability in the selected part of the deposit. There is an apparent narrowing of values in the ordinary Kriging model. Note that the Kriging smoothing effect can be compensated by implementing the Yamamoto correction.

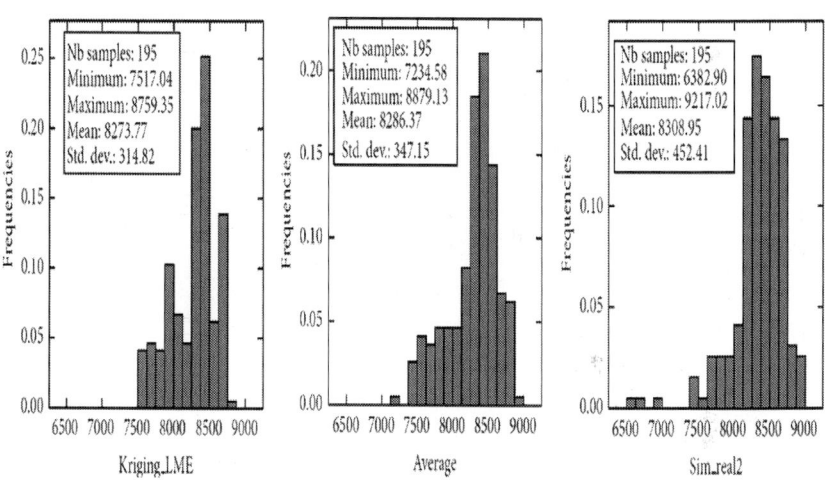

Figure 6: Histograms of calorific value Q_i^r based on ordinary Kriging model (l), average of 50 realizations of a geostatistical simulation (m), and one realization of a geostatistical simulation (r).

Figure 7(a) shows the standard deviation of ordinary Kriging, which expresses the magnitude of the expected interpolation error.

Its size in any given block depends primarily on the distance to the nearest observation, on the basis of which the interpolation was conducted. This relation results mainly from the variogram and the data configuration.

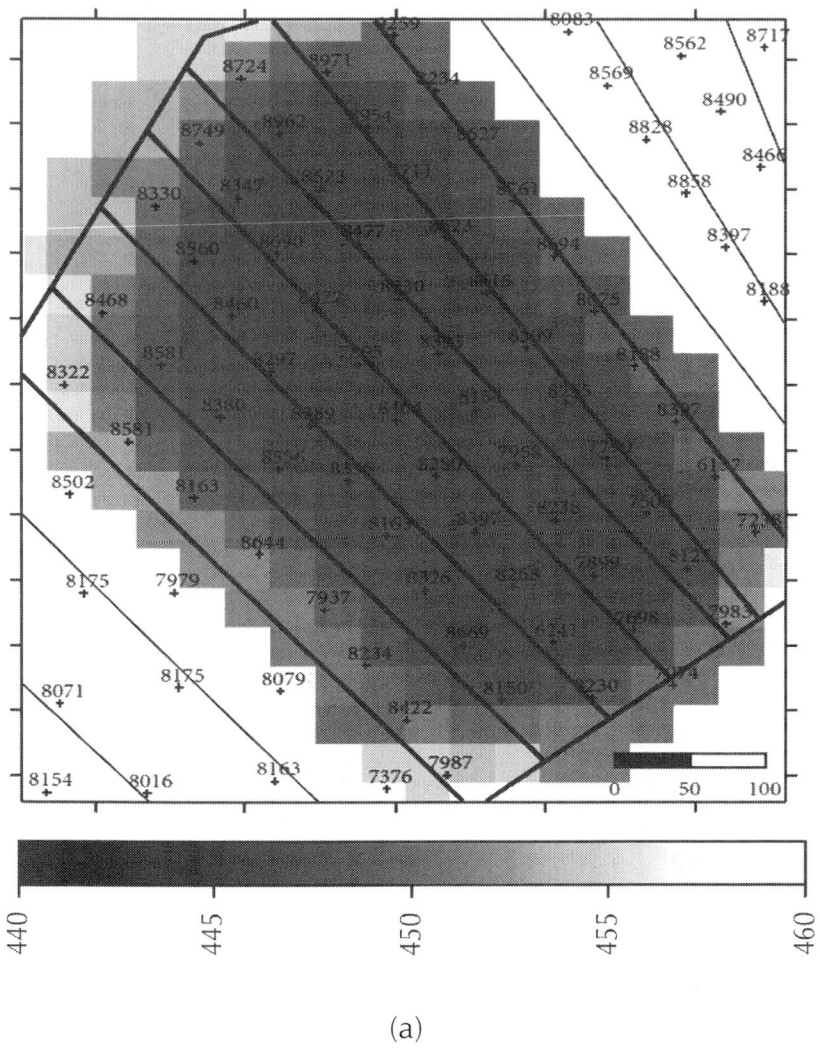

(a)

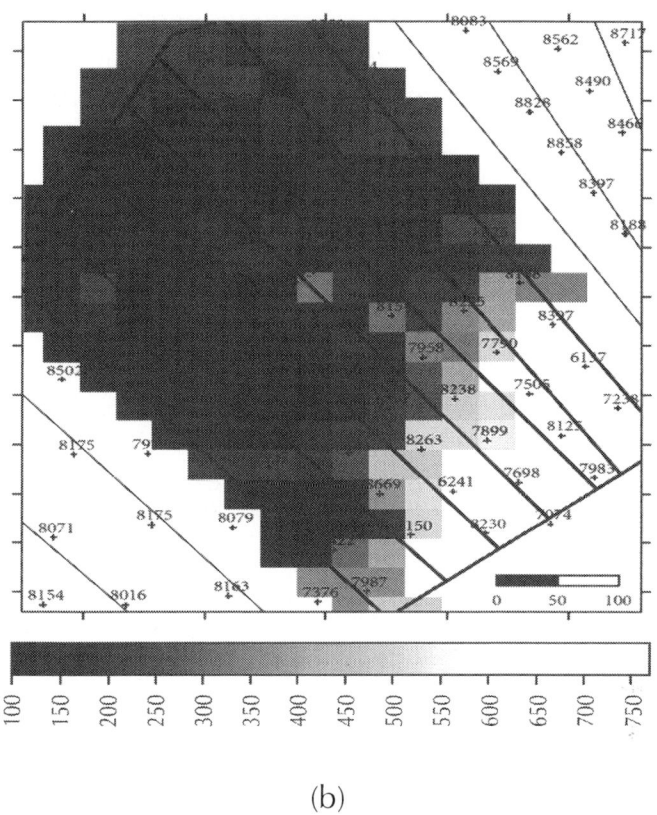

(b)

Figure 7: Standard deviation of Kriging (l) and standard deviation of simulation based on 50 realizations (r).

Kriging's standard deviation is independent of local variation of observations used for modelling. Figure 7(b)presents the map of the conditional simulation's standard deviation. The map is a result of a statistical analysis of 50 realizations. In each node of the grid standard deviation was calculated, reflecting the uncertainty of a local forecast. There are some clear differences between the two figures. These differences appear not only in the nominal value of the standard deviation, but also in their spatial distribution in the modelled deposit. The standard deviation of the simulation shows particularly high values in the south-eastern part of the deposit. This is the influence of high calorific value variation of the adjacent observations.

Based on the assumed extraction schedule (Figure 2), graphs of the calorific value in the subsequently mined blocks (corresponding to the average daily production volumes) were prepared. Figures 8 and 9 depict the variations of the calorific value in the lignite stream during six months of mining. The graph in Figure 8 was created using the variability model prepared with the use of the ordinary Kriging method. Besides the mean value, the dotted lines constitute for the Kriging's standard deviation of the respective exploitation blocks. The graph in Figure 9 shows the variation of the calorific value based on the model created by conditional simulation method. Three exemplifying realizations of the simulation are shown together with the mean of all 50 realizations. Graphs (Figures 8 and 9) are supplemented with horizontal lines corresponding to the average

value calculated from 68 observations (Q_{sr} = 8267 kJ / kg) and the lines corresponding to the average increased (Q_g=8788kJ/kg) and the average decreased by the value of the standard deviation of the observation (Q_d = 7746kJ/kg).

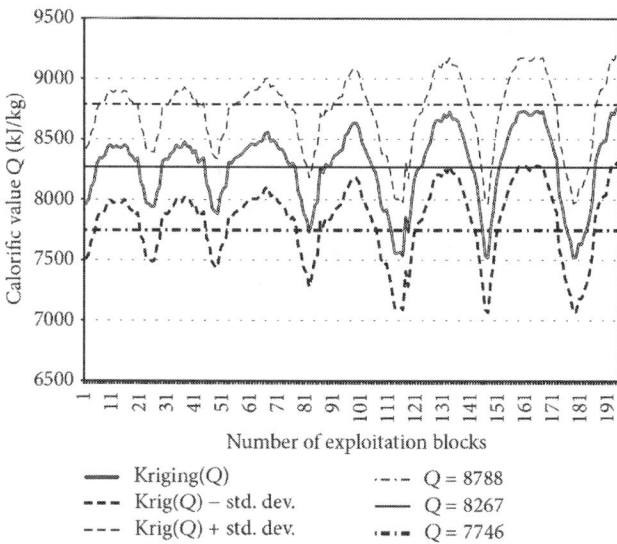

Figure 8: The fluctuation of calorific value Q_i^r within 195 days of exploitation—based on ordinary Kriging.

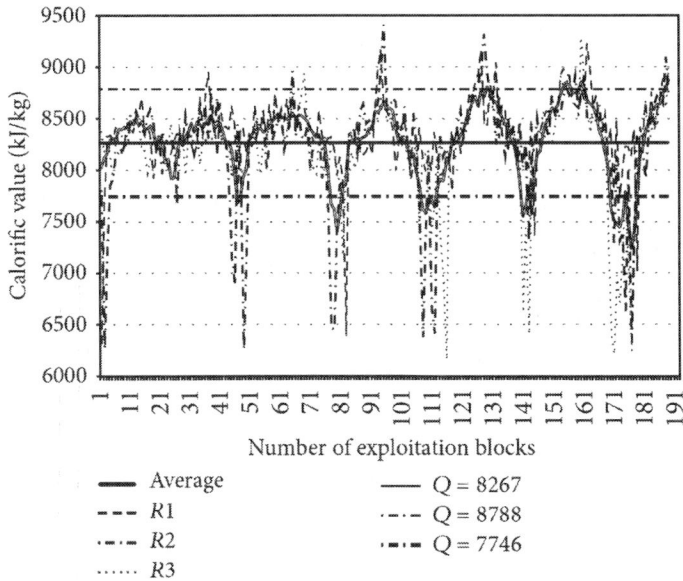

Figure 9: The fluctuation of calorific value Q_i^r within 195 days of exploitation—based on conditional simulation.

In the first graph (Figure 8), the average calorific value determined using Kriging changes cyclically within the range of the standard deviation of observation, extending only slightly beyond those lines. In the second graph (Figure 9) the volatility of individual realization (R1, R2, and R3) is significant, and the mean of realizations (average R1, R1, ..., R50) in several places goes far beyond the limits of the line marking Q_d=7746 kJ/kg, reaching a value below Q=6500kJ/kg.

Based on the results of the 50 realizations of simulation, a map showing the probability of exceeding the thresholds established in the particular exploitation blocks was created. The values of the mean of 68 observations plus and minus the standard deviation were chosen as the assumed limits (thresholds), which is rounded, respectively, to Q_g=8790 kJ/kg and Q_d=7750 kJ/kg. Figure 10 shows the map of the probability of exceeding the adopted thresholds.

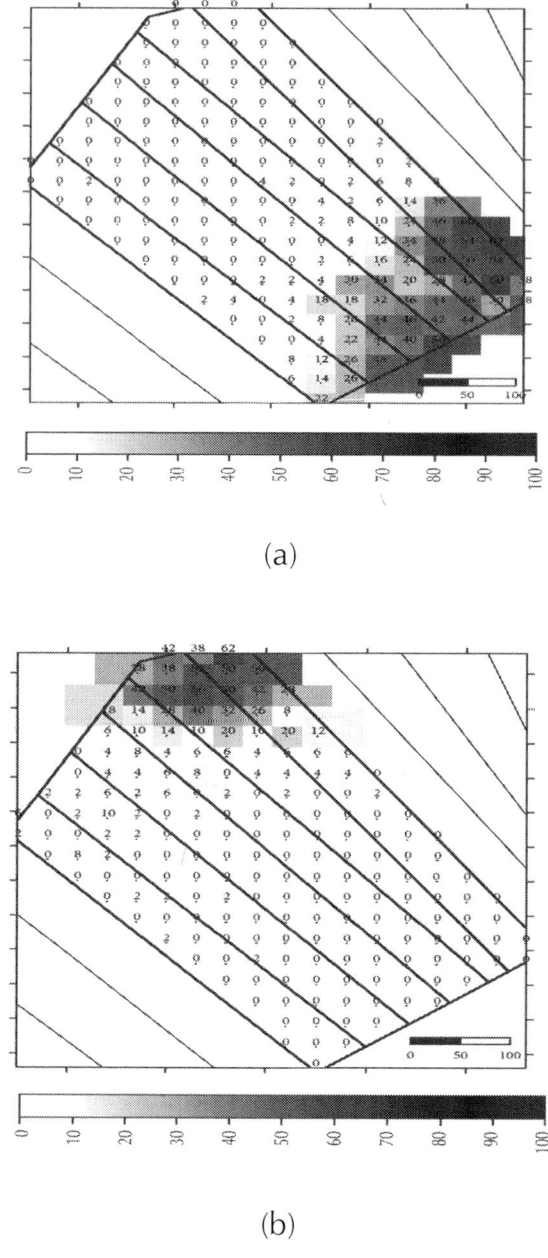

(a)

(b)

Figure 10: The occurrence probability of calorific value lower than Q_d=7750 kJ/kg (l) and higher than Q_g=8790 kJ/kg (r) in the exploitation blocks, based on simulation.

INVESTIGATION OF DESIGN OPTIONS

Option 1 (Bed Blending using a Coal-stock-and-blending Yard)

Bed blending has three objectives: namely, buffering, composing, and homogenising. Thereby it transforms the characteristics of the incoming material flow in an outgoing material flow, whose characteristics are defined by costumer specifications and may be of contractual relevance. The characteristics of the incoming material flow are a function of the geological conditions, the applied selectivity in extracting the deposit, the mining sequence, and the operation mode in the pit as discussed in the previous section. The following considerations concern the homogenisation effect of using bed blending. The efficiency of blending and smoothing variability is significantly dependent on constructive parameters as well as the operation of the blending yard. Constructive factors are the type of the yard, its length and width, the angle of repose, the number of layers, and speed of the stacker. The following constructive parameters are given in the case study: the blending yard is of type "strata" (Figure 11).

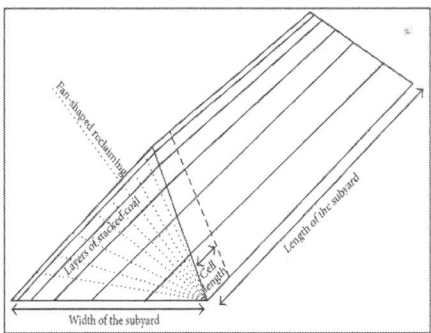

Figure 11: Schematic illustration of a strata-type blending yard.

The coal coming from the pit (incoming material flow) is stacked into layers, which are spread along the total length of a bed by a continuously up and down moving stacker. The number and thickness of the several layers are variable and can be influenced by the moving rate of the stacker dependent on the total production rate of the mine. At maximum about 61 layers can be placed in a pocket. The yard is reclaimed in a fan-shaped manner orthogonally to the alignment of the stacked layers by a scraper. In this way the coal quality of the outgoing material flow is formed as an average over the total number of stacked layers.

Investigations have shown that operating with >15 layers the incoming flow can be completely homogenized [5, 12]. Therefore in this investigation it is assumed that the homogenisation effect is solely dependent on the stockpile size. Figure 12 shows the variability of the outgoing material flow for the different blending yard sizes: 0 kt, 60 kt, 180 kt, and 300 kt. Clearly already a considerable small blending bed size leads to a significant homogenization. Considering the already previously introduced lower and upper limit of Q_g=8790kJ/kg and Q_d=7750kJ/kg it would need a stockpile size of >300 kt to ensure continuously in-spec delivery of the power plant.

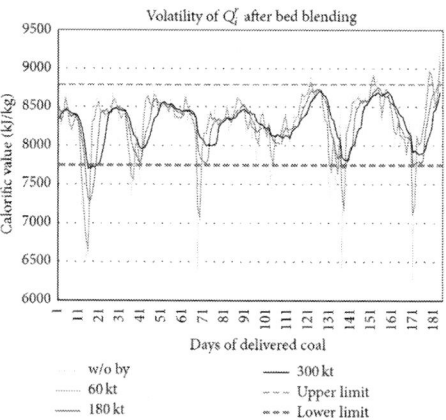

Figure 12: The fluctuation of calorific value Q_i^r within 195 days of exploitation—after bed blending.

Figure 13 shows a summary of the frequencies of expected deviations from production targets for different blending bed sizes. For example a size of 180 kt would still lead to approximately 5% of daily deliveries deviating from potentially contractually fixed limits. A size of 330 kt would ensure that the in situ variability of the deposit can be transformed into a product exhibiting a maximum variability as requested from the customer. In addition this size of a stock pile would form a buffer bridging about 11 days of production and can ensure continuous supply of the power plant during small and medium termed maintenance or breakdown events.

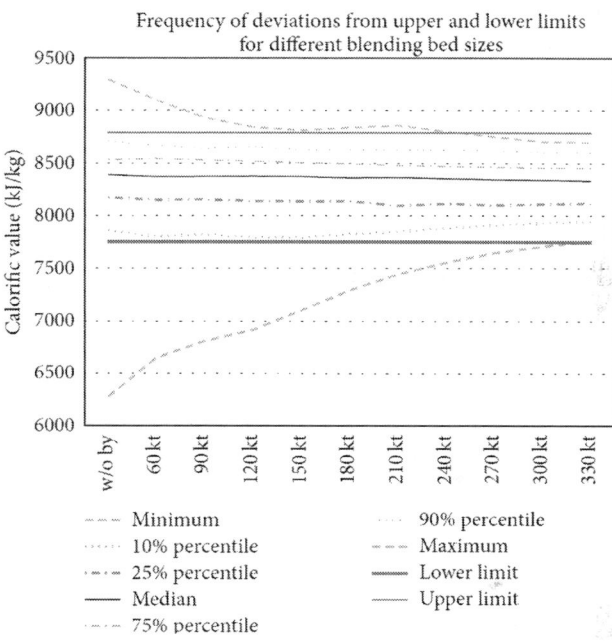

Figure 13: Distribution of calorific value as a function of blending yard sizes.

Option 2 (Availability of Two Excavators)

This design option considers the availability of two excavators, which are operated simultaneously. For example excavator one may excavate the first part of the bench to the middle and excavator two extracts the remaining blocks. To avoid installed overcapacity the capacity of each of the two excavators can be designed as low as half of the capacity of on single excavator achieving the same daily production target of 30 kt. For this investigation it was assumed that both excavators operate at an extraction rate of 15 kt per day. Assuming no targeted quality optimized scheduling, which means both excavators are always operating at one half of the bench without pinpointed schedule, Figure 14 shows the result of the blended stream of lignite. As can be seen, a simultaneous extraction of blocks with a subsequent blending on the belt conveyor significantly reduces the variability. Considering the coal quality production limits it becomes obvious that there may still occur sporadic deviations from production targets. These can be avoided by quality optimized scheduling or using an additional blending yard with a small capacity, for example, 60 kt.

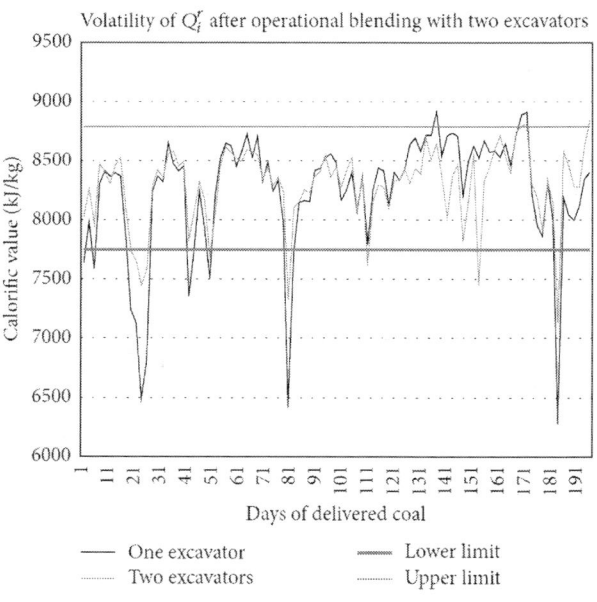

Figure 14: The fluctuation of calorific value Q_i^r within 195 days of exploitation—using two excavators simultaneously.

CONCLUSIONS

The calorific value of the analysed part of the deposit has a relatively low volatility (6,3%), yet due to the elongated shape of the deposit (Figure 1), which implies the direction of mining and distribution of the calorific value, the average daily calorific values are in the range of 7750–8790 kJ/kg.

With the accepted method of mining, changes occur in almost regular monthly cycles. In two parts of the deposit the lignite has wider-than-threshold values (Figure 10). This applies mainly to the values lower than 7750 kJ/kg in the south-eastern area, as well as more than 8790 kJ/kg in the northern part of the deposit. For purposes of coal quality control in order to maintain the calorific value at the desired level, it is useful to apply modern spatial interpolation tools. The study shows that for this purpose geostatistical simulation is particularly useful as it—in addition to the mean values—allows determining the level of the probability of exceeding the adopted thresholds in the particular blocks (risk level). In contrast to the simulation, using the ordinary Kriging interpolation may lead to erroneous operating decisions because of the effect of smoothing of the extreme values demonstrated in the paper.

The property of geostatistical simulation to reproduce in situ variability can be used to investigate the variability in dependence of certain design options in the subsequent material handling system. In the present case the availability and size of blending beds were investigated as well as the availability of an additional excavator. It has been shown that both options can contribute significantly to the reduction of variability in CV. In addition a required stock-pile size could be defined that ensures a continuous in-spec delivery of coal to the customer.

REFERENCES

1.   W. Naworyta and S. Sypniowski, "About the problem of lignite stream quality control in the context of proper identification

of deposit's quality parameters," Surface Mining, no. 2, pp. 58–65, 2013 (Polish).

2. P. M. Gy, "A new theory of bed-blending derived from the theory of sampling—development and full-scale experimental check," International Journal of Mineral Processing, vol. 8, no. 3, pp. 201–238, 1981. · ·

3. M. Kumral, "Bed blending design incorporating multiple regression modelling and genetic algorithms,"The Journal of the Southern African Institute of Mining and Metallurgy, vol. 106, no. 3, pp. 229–236, 2006. ·

4. D. Marques, J. F. Costa, D. Ribeiro, and J. C. Koppe, "The evidence of volume variance relationship in blending and homogenisation piles using stochastic simulation," in Proceedings of the 4th World Forum on Sampling and Blending, pp. 235–242, The Southern African Institute of Mining and Metallurgy, 2009.

5. J. Benndorf, "Investigating the variability of key coal quality parameters in continuous mining operations when using stockpiles," in Advances in Orebody Modelling and Strategic Mine Planning I, AusIMM, 2011.

6. J. Benndorf, "Investigating in situ variability and homogenisation of key quality parameters in continuous mining operations," Transactions of the Institutions of Mining and Metallurgy, Section A: Mining Technology, vol. 122, no. 2, pp. 78–85, 2013. · ·

7. D. Gärtner and R. Hempel, Monitoring and Control of Processes in the Lignite Mines in Rhineland, Lignite Mining, Springer, Heidelberg, Germany, 2009.

8. L. Kunde and D. Trummer, "Coal quality management, lignite mining—Kohlenqualitätsmanagement (germ.)," in Der Braunkohlentagebau, pp. 409–426, Springer, Berlin, Germany, 2009.

9. B. Zimmer, "Development of a new on-line coal quality management system in a lignite mine in Serbia," in Continuous Surface Mining, Latest Development in Mine Planning,

Equipment and Environmental Protection: Proceedings of 10th International Symposium Continuous Surface Mining, 13–15 September 2010, C. Drebenstedt, Ed., pp. 290–302, Technische Universitat Bergakademie Freiberg, Freiberg, Germany, 2010.

10.   W. Naworyta, "Variability analysis of lignite deposit parameters for output quality control," Mineral Resources Management, vol. 24, no. 2–4, pp. 97–110, 2008 (Polish).

11.   W. Naworyta and J. Benndorf, "Accuracy assessment of geostatisticalmodelling methods of mineral deposits for the purpose of their future exploitation—based on one lignite deposit," Mineral Resources Management, vol. 28, no. 1, pp. 77–101, 2012 (Polish).

12.   J. Benndorf, "Application of efficient methods of conditional simulation for optimising coal blending strategies in large continuous open pit mining operations," International Journal of Coal Geology, vol. 112, pp. 141–153, 2013. ·

13.   N. Remy, A. Boucher, and J. Wu, Applied Geostatistics with S-GeMS, Cambridge University Press, Cambridge, UK, 2009.

Assessment of Cutting Parameters Influencing on Thrust Force and Torque during Drilling Particulate Filled Glass Fabric Reinforced Epoxy Composites

Bhadrabasol Revappa Raju[1], Bheemappa Suresha[2], Ragera Parameshwarappa Swamy[3], and Bannangadi Swamy Gowda Kanthraju[4]

[1]Department of Mechanical Engineering, PES Institute of Technology and Management, Shivamogga, India

[2]Department of Mechanical Engineering, The National Institute of Engineering, Mysore, India

[3]Department of Mechanical Engineering, University B.D.T. College of Engineering, Davangere, India

[4]Department of Mechanical Engineering, Don Bosco Institute of Technology, Bangalore, India

ABSTRACT

Drilling is indispensable process and it cannot be avoided for joining composite structures used in various engineering applications. In this research article, the influence of drilling parameters on thrust force and torque of silica (SiO_2) and alumina (Al_2O_3) filled into glass fabric reinforced epoxy (G-E) composites are analyzed. Drilling experiments are conducted on these composite materials using BATLIBOI make radial drilling machine. Two different drill bits (HSS and cemented carbide) are used for the experimentation. The influence of drilling parameters like cutting speed and feed on thrust force and torque on drilling of particulate filled G-E composites has been carried out. The experimental results indicated that the thrust force and torque were increased with increasing feed and cutting speed for all the composites tested. Further, it is observed that the carbide drill performed better than HSS drill during drilling of particulate filled G-E composites. The drilled surfaces are examined using scanning electron microscopy (SEM) and damage mechanisms are discussed.

INTRODUCTION

The use of fiber reinforced composite materials has grown in recent years in every field of engineering due to their inherent advantages over conventional materials. However, due to the presence of two or more dissimilar phases, composite materials expose challenges during machining as well as material characterization. Polymer based composites also provide good design, flexibility and high dielectric strength and usually require lower tooling costs. In machining processes, however, the quality of the component is greatly influenced by various parameters such as cutting conditions, tool geometry, tool material, machining process, chip formation, workpiece material, tool wear, vibration during cutting, etc. Drilling which is a secondary machining process is performed on fiber reinforced polymer composites (FRPCs) to know the effect of various factors during machining. Drilling is the most widely used for machining of composite materials, the effect of thrust force and torque that occurs during drilling. Therefore, process for composites which are used in aerospace, automotive and machine tool industries. The main concern in the drilling of composite it is necessary to understand the relationship among the various controllable parameters and to identify the important parameters that influence the quality of holes drilled.

Mathew et al. [1] studied that thrust is a major factor responsible for delamination and it mainly depends on tool geometry and feed rate. Trepanning tools, which were used in this study, were found to give reduced thrust while making holes in thin laminated carbon fiber reinforced polymer composites (CFRPCs). In his work the peculiarities of trepanning over the drilling of unidirectional composites has been emphasized. Sonbaty et al. [2] studied the influence of some parameters on the thrust force, torque and surface roughness in drilling processes of FRP. These parameters include cutting speed, feed, drill size and fiber volume fraction. Zitoune et al. [3] studied the parametric influences on thrust force, torque as well as surface finish, on CFRPCs. The experimental results showed that the quality of holes can be improved by proper selection of cutting parameters. This is substantiated by monitoring thrust force, torque, surface finish, circularity and hole diameter. Capello and Tagliaferri [4] studied to clarify the interaction mechanisms between the drilling tool and material. Drilling tests were carried out on glass-polyester composites using standard HSS tool, drilling was interrupted at preset depths to study damage development during drilling. Arul et al. [5] suggested that the drilling of polymeric composites which aimed to establish a technology that would ensure minimum defects and longer tool life using HSS drill. Using HSS drill, a series of vibratory drilling and conventional drilling experiments were conducted on glass fabric reinforced polymer composites (GFRPCs) to determine delamination factor. Hocheng and Tsao [6] studied the critical thrust force and delamination on CFRPCs and compared the effects of these on different drill bits. The advantage of these special drills were illustrated mathematically as well as experimentally, that their thrust force is distributed toward the drill periphery instead of being concentrated at the center. Lin and Chen [7] carried out a study on drilling composite materials at high speed and concluded that an increase in the cutting velocity leads to an increasing tool wear that in turn provokes it an increase in the thrust force. They studied the effects of increasing cutting speed on drilling characteristics of CFRPCs. The effects of increasing cutting speed range from 9550 to 38,650 rpm on average thrust force, torque, tool wear and hole quality for both multifaceted drill and twist drill are studied.

Khashaba et al. [8] reported that the delamination-free in drilling different fiber reinforced thermoset composites was the main objective of research. Therefore the influence of drilling and material variables on thrust force, torque and delamination of GFRP composites were investigated experimentally using multifaceted and twist drill. Piquet et al. [9] suggested that the effect of thrust on drilling with a twist drill and a specific cutting tool of structural thin backing plates in carbon-epoxy. The possibility to manufacture carbon/epoxy with a conventional cutting tool was analyzed and the limits of the twist drill were shown. Krishnaraj et al. [10] studied the damage generated during the drilling of GFRPCs which was detrimental for the mechanical behavior of the composite structure. The work was focused on analyzing the influence of drilling parameters (spindle speed and feed) on the strength of the woven glass fabric reinforced polymer laminates and further to study the residual stress distribution around the hole after drilling.

Abrao et al. [11] conducted the review of drilling of fiber reinforced plastics. This review highlighted the various aspects of on drilling of glass and carbon fiber reinforced polymers. Further, aspects such as tool materials and geometry, machining parameters and their influence on the thrust force and torque are investigated.

Additionally, the quality of the holes produced is also assessed, with special attention paid to the delamination damage.

Tsao et al. [12] investigated that the experimental results indicate that the feed rate and the drill diameter are the most significant factors affecting the thrust force, while the feed rate and spindle speed contribute the most to the surface roughness. In this study, the objective was to establish a correlation between the feed rate, spindle speed and drill diameter with the induced thrust force and surface roughness in drilling composite laminate.

Krishnaraj et al. [13] investigated that an experimental investigation has been performed on GFRPCs using carbide drill with different drill geometries, namely standard twist drill, double cone drill, Zhirov-point drill and multifacet drill. A series of experiments are conducted using a wide range of cutting parameters namely, speed and feed rate. Thrust force and surface roughness are measured and studied in the test trials. The relation between spindle speed and feed rate on thrust force and surface roughness is established. It is found that Zhirov-point drill and multifacet drill could be used at high spindle speed which generates less thrust force.

Malhotra et al. [14] studied that the drilling of glass fiber/epoxy and carbon fiber/epoxy laminates is studied using HSS and tungsten carbide coated drills. The effect of cutting speed, feed and the number of holes on tool wear, thrust and torque is studied. Carbide drill performs much better than HSS drill with both materials. Thrust force and torque are much higher in the drilling of CFRPCs as compared to GFRP. Ramkumar et al. [15] studied that coated HSS drill performs a little better than uncoated HSS for a small number of holes, while their performance is inferior to uncoated HSS for a large number of holes.

Based on the thorough literature cited above, the present work aimed at ascertaining the effects of drilling parameters on thrust force and torque of woven glass fabric reinforced epoxy (G-E) composites filled with ceramic fillers by using two different drill bits namely two flute HSS and cemented carbide.

EXPERIMENTAL DETAILS

Materials and Fabrication

The matrix material system selected is an Epoxy resin (LAPOX L-12 with density 1.16 g/cm³) supplied by ATUL India Ltd., Gujarat, India. Woven glass plain weave fabrics made of 360 g/m², containing E-glass fibers of diameter of about 12 μm have been used as the reinforcing material in all the composites. The fillers chosen were Silicon dioxide (SiO_2) and aluminum oxide (Al_2O_3). The average particle size of SiO_2 and Al_2O_3 micro particles are about 10 μm size. The details of the compositions are listed in Table 1. As regards to the processing, on a Teflon sheet, E-glass woven fabric was placed over which the epoxy matrix system consisting of epoxy and hardener was smeared. Dry hand lay-up technique was employed to fabricate the composites. The stacking procedure consists of placing the fabric one above the other with the resin mix well spread between the fabrics. A porous Teflon film was again used to complete the stack. To ensure uniform thickness of the sample, a 10 mm spacer was used. The mould plates were coated with release agent in order to aid the ease of separation on curing. The cast of each composite after 12 h of impregnation and dried for 2 h at 100°C followed by compression molding at a temperature of 390°C and a pressure of 7.35 MPa. The slabs so prepared measured 250 mm × 250 mm × 10 mm in size.

To prepare different wt% of particulate filled G-E composites, besides the epoxy hardener mixture, additional particulates were included to form the resin mix. The details of the composites selected for the study are listed in Table 1. The percentage of the glass fiber in the composite is 60 by wt%.

Physico-Mechanical Tests

The density of the composites was determined by using a high precision electronic balance (Mettler Toledo, Model AX 205) using the Archimedes principle and using Durohardness tester, the hardness of the composites is measured, the values recorded. Tensile properties were measured using a Universal testing machine in accordance with the ASTM D-3039 procedure at a cross head speed of 5 mm/min and a gauge length of 50 mm. The tensile strength and modulus were determined from the stressstrain curves. Five samples were tested in each set and the average value was reported. The tensile test was carried out on a fully automated Lloyd LR-20 kN Universal testing machine connected to a computer with DAPMAT software. The Physico-mechanical properties of epoxy resin, glass fiber and fillers are shown in Table 2.

Machining Set-Up and Drilling Procedure

Drilling was performed on a BATLIBOI make radial drilling machine (Figure 1) by using HSS and Carbide twist drill of 6 mm diameter with 118° point angle. The cutting force components such as thrust force F (N) and torque T (N-m) for different cutting conditions were measured using drill tool dynamometer. Drilling operations were carried out on composite laminates in a dry environment with the cutting conditions are listed in Table 3. Components of cutting force i.e., thrust and torque were monitored at regular intervals of every 10 holes machined using a real time data acquisition system. Worn out surfaces and drilled surfaces are examined using scanning electron microscope (SEM) and analyzed.

Table 1: Compositions of particulate filled glass-epoxy composites

Sl. No	Composite	Matrix (wt%)	Reinforcement (wt%)	Filler (wt%)
1	Glass-epoxy	Epoxy (40)	Glass fabric (60)	
2	Silicon dioxide-glass-epoxy	Epoxy (30)	Glass fabric (60)	SiO_2 (10)
3	Alumina-glass-epoxy	Epoxy (30)	Glass fabric (60)	Al_2O_3 (10)

Table 2: Physico-mechanical properties of epoxy resin, glass fiber and fillers

Property	Epoxy resin	Glass fibers	Silicon dioxide (filler)	Alumina (filler)
Density (g/cm³)	1.16	2.54	2.2	3.89
Tensile strength (MPa)	110	3400	110	260-300
Tensile Modulus (GPa)	4.1	72.3	73	375

Table 3: Cutting conditions employed in the present study

Spindle speed (m/min.)	15.08	18.85	23.57	30.16
Feed (mm/rev.)	0.18	0.36	0.71	1.4
Diameter of drill bit used (mm)	6	6	6	6
Cutting medium	Dry	Dry	Dry	Dry

Figure 1: Drilling machine showing the dynamometer used for measurement of cutting forces.

RESULTS AND DISCUSSION

Evaluation of Physico-Mechanical Properties

The physico-mechanical properties such as density, hardness, tensile strength and tensile modulus data of unfilled and particulate filled G-E composites are given in Table 4. The results revealed that particulate filled GE composites showed better mechanical properties than that of unfilled G-E composites. Comparing the results it was observed that the inclusion of ceramic fillers into G-E showed higher density. The incorporation of SiO_2 and Al_2O_3 filler in G-E composites increased the tensile strength. Elongation properties decreased with the presence of filler that indicates interference by the filler in the deformability of the matrix. Al_2O_3 filled G-E composites showed improved mechanical properties compared to unfilled and SiO_2 filled G-E composites. It should be pointed out that the presence of SiO_2 and Al_2O_3 fillers improved adhesion and it has been proved to be beneficial in glass fiber reinforced epoxy composites.

Effect of Cutting Speed and Feed on Thrust and Torque Using HSS Drill Bit

Figures 2(a) and (b), and Figures 3(a) and (b) show the parametric variation of thrust force and torque with feed rate for two cutting speeds of particulate filled G-E composites using HSS tool.

It can be observed that thrust force and torque generally increased with feed rate. This fact was due to the increasing the cross-sectional area of the undeformed chip.

Table 4: Physico-mechanical properties of G-E, SiO_2 and Al_2O_3 filled G-E composites

Sample code Properties	G-E	10% SiO_2-G-E	10% Al_2O_3-G-E
Density (g/cm³)	1.98	2.19	2.30
Hardness (Shore-D)	63	66	72
Tensile strength, σ (MPa)	254	326.7	352
Tensile modulus, E (GPa)	8.34	9.57	11.55

The presence of ceramic filler in composites increases the hardness and cutting resistance of the material, also may result in wears the cutting edges of the drill through drilling one hole. Therefore the thrust force and torque were increased with increasing cutting speed. The addition of filler reduces the thrust force for both 15.08 and 30.16 m/min cutting speeds. From Figures 1 and 2, it can be seen that alumina filled G-E composites showed the lower thrust and torque value for different cutting speeds as compared to the unfilled and SiO_2 filled G-E composites.

Effect of Cutting Speed and Feed on Thrust and Torque Using Carbide Drill Bit

Figures 4(a) and (b), and Figures 5(a) and (b) show the influence of drilling variables on peak thrust force and torque, respectively, for unfilled and particulate filled composites. The results in these figures indicate that, the thrust force and torque were increased with increasing feed. This fact was due to the increasing the cross-sectional area of the undeformed chip. The presence of fillers with the abrasive nature in G-E composite that, in addition to increasing the hardness and cutting resistance of the material, also may result in wears the cutting edges of the drill through drilling one hole. Therefore the thrust force and torque were increased with increasing cutting speed (Figures 4 and 5).

Drilled Surface Morphology for HSS Drill Bit

The SEM micrographs show the breakage of the fiber material and damage of the matrix material.Figure 6(a) shows the less breakage of fiber material and also ruptured matrix but as the speed and feed are increasing the breakage of fibers is more as seen and also more ploughing and fiber buckling of the fibers with adhesion of matrix debris on the fiber surface as shown in Figure 6(b). The SEM micrographs show the breakage of the fiber material and damage of the matrix material.Figure 7(a) shows the less breakage of fiber material and also damaged matrix but as the speed and feed are increased the breakage of fibers is more as seen in Figure 7(b) due to delamination between the layers of the material. At higher cutting conditions (high speed and feed), the thrust force caused visible cracking of surface layer as seen in Figure 7(b) which resulted in deterioration of the surface, due to high speed and feed which includes fiber fragmentation, matrix debris, large number of fiber breakage, inclined fiber fracture, severe debonding at fiber matrix interface.

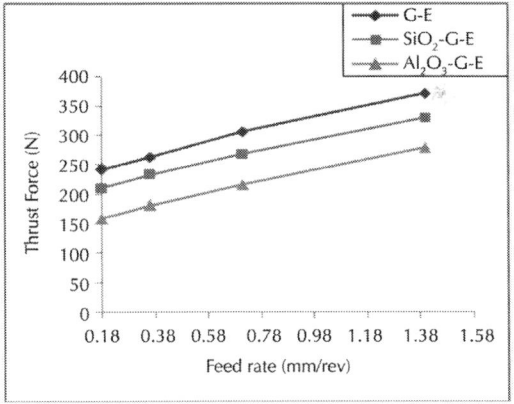

(a)

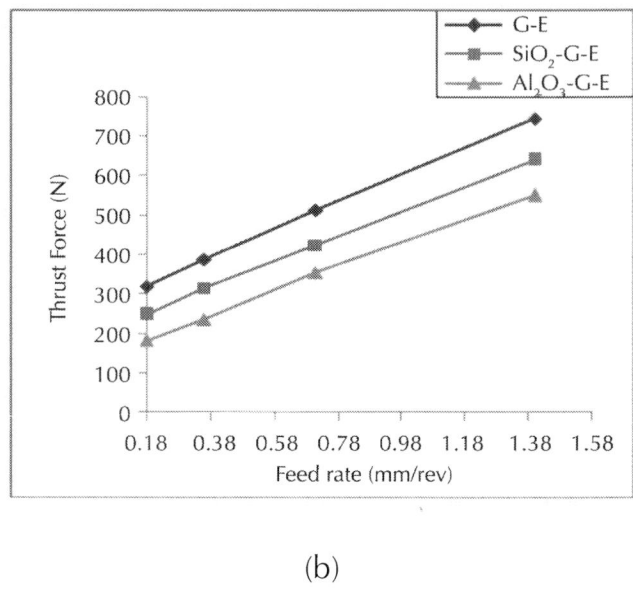

(b)

Figure 2: Variation of thrust force with feed rate for filled & unfilled G-E composites using HSS drill bit at (a) 15.08 m/min and (b) 30.16 m/min.

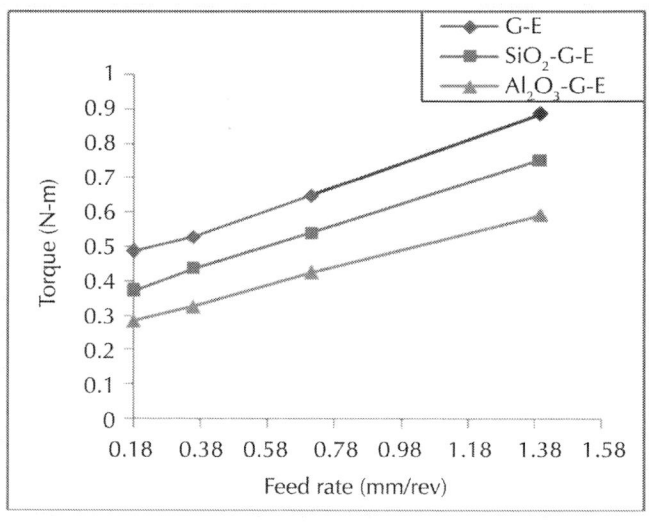

(a)

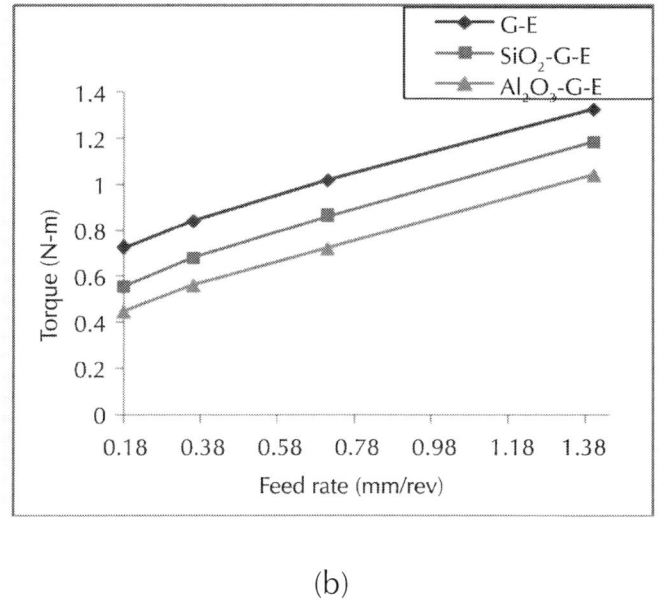

(b)

Figure 3: Variation of torque with feed rate for filled and unfilled G-E composites using HSS drill bit at (a) 15.08 m/min and (b) 30.16 m/min.

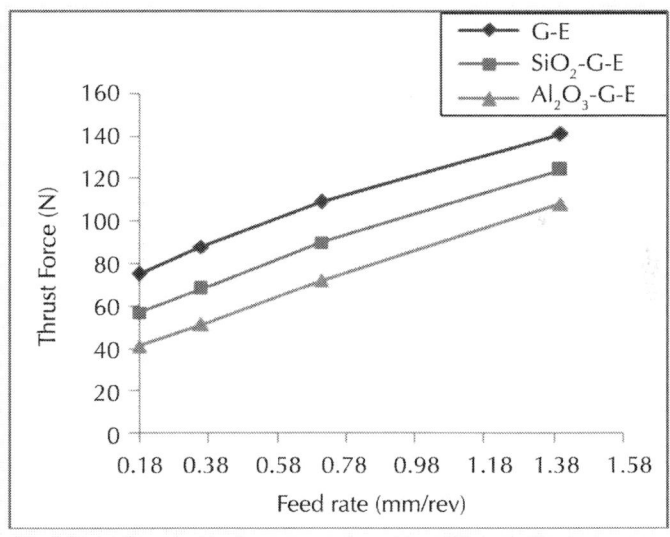

(a)

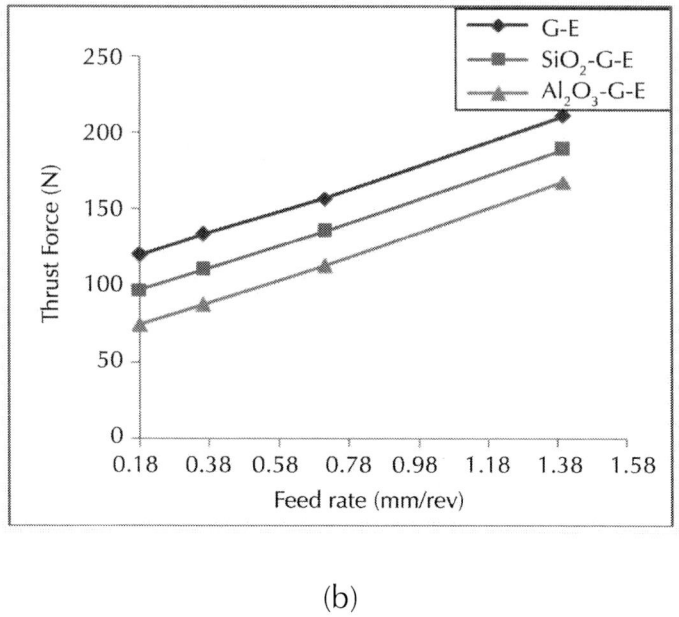

(b)

Figure 4: Variation of thrust force with feed rate for filled and unfilled G-E composites using carbide drill bit at (a) 15.08 m/min and (b) 30.16 m/min.

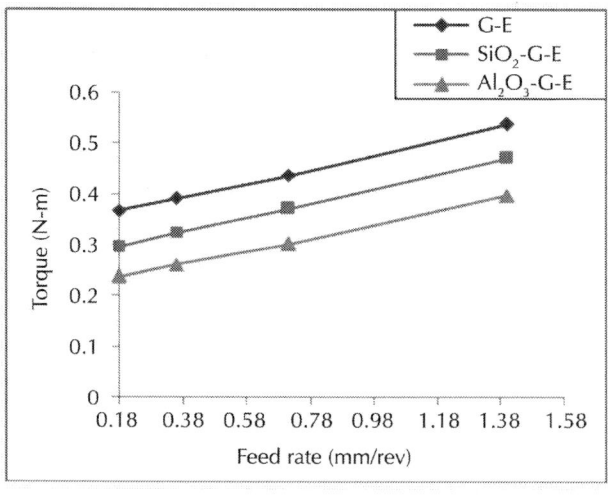

(a)

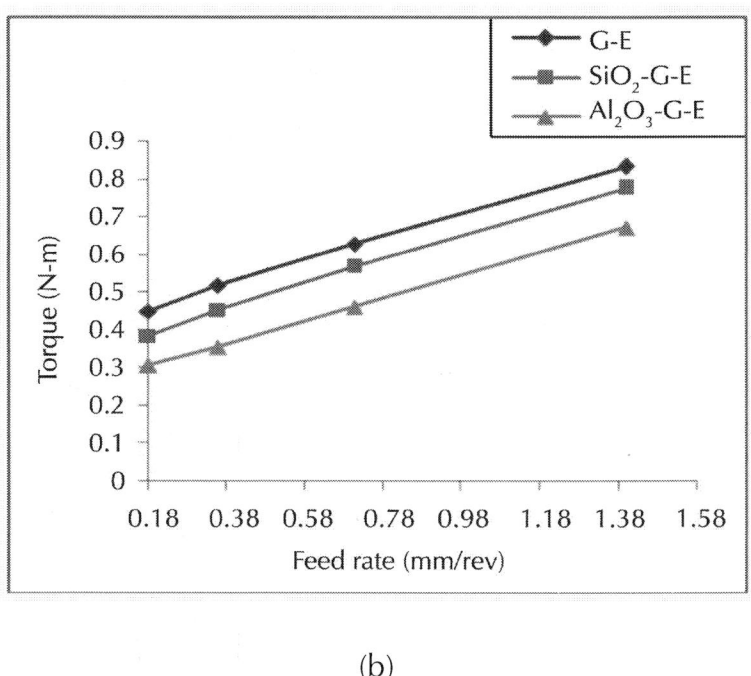

(b)

Figure 5: Variation of torque with feed rate for filled and unfilled G-E composites using carbide drill bit at (a) 15.08 m/min and (b) 30.16 m/min.

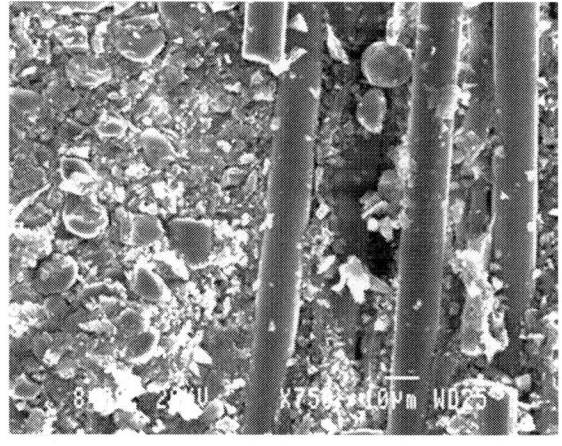

(a)

(b)

Figure 6: SEM Micrographs of the G-E using HSS drill bit at (a) 15.08 m/min, 0.18 mm/rev and (b) 30.16 m/min, 1.40 mm/rev.

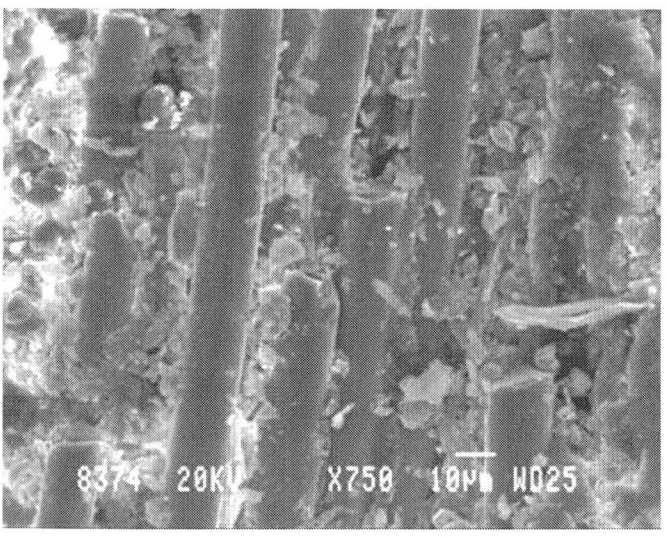

(a)

(b)

Figure 7: SEM Micrographs of the G-E with SiO_2 using HSS drill bit at (a) 15.08 m/min, 0.18 mm/rev and (b) 30.16 m/min, 1.40 mm/rev.

The SEM micrographs show the breakage of the fiber material and damage of the matrix material.Figure 8(a) shows the fibers are totally misaligned and the formation of voids in between the matrix and the fiber is clearly observed and the drilled surface at lower speed and feed seems to have less no of cracks resulting in lower value of surface roughness. In the Figure 8(b) it shows the breakage of fiber material and also the ploughing action of fibers due to the poor bonding between fibers and matrix, inclined fiber breakage and cohesive resin fracture.

Drilled Surface Morphology for Carbide Drill Bit

The SEM micrographs show the breakage of the fiber material and damage of the matrix material.Figure 9(a) shows that the longitudinal fibers are pulled up and are oriented and more breakage of matrix material such as cohesive bonding of matrix and fiber at interface, more transverse fiber breakage, matrix debris, voids and fiber fragmentation but as the speed and feed are increased the breakage of fibers as well as the matrix is increased and also chopping of fibers is being seen inFigure 9(b).

Figure 10(a) shows the less breakage of fiber material and more damage to the matrix material at lower speed and feed the drilled work surface showed the similar surface features such as cohesive bonding of matrix and fiber at interface, more transverse fiber breakage, matrix debris, voids and fiber fragmentation and as the speed and feed are increased the e breakage of fibers and matrix is more as seen in Figure 10(b).

(a)

(b)

Figure 8: SEM micrographs of the G-E material with Al_2O_3 filler using HSS drill bit at (a) 15.08 m/min, 0.18 mm/rev and (b) 30.16 m/min, 1.40 mm/rev.

(a)

(b)

Figure 9: SEM micrographs of the G-E using carbide drill bit at (a) 15.08 m/min, 0.18 mm/rev and (b) 30.16 m/min, 1.40 mm/rev.

(a)

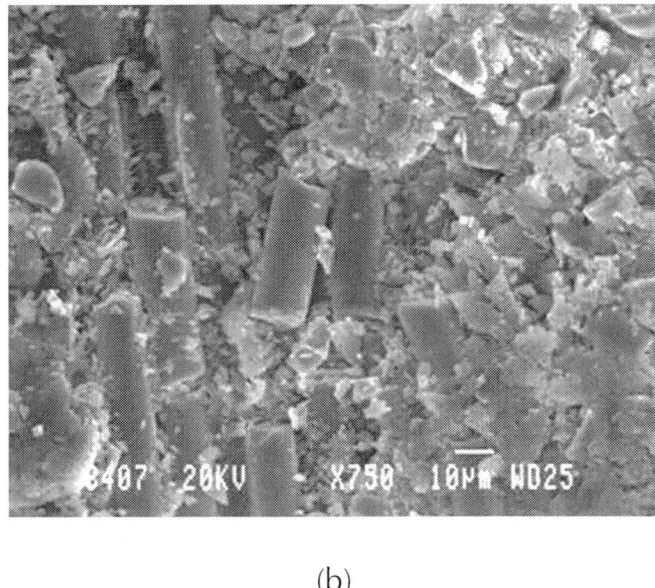

(b)

Figure 10: SEM micrographs of the G-E with SiO_2 using carbide drill bit at (a) 15.08 m/min, 0.18 mm/rev and (b) 30.16 m/min, 1.40 mm/rev.

(a)

(b)

Figure 11: SEM Micrographs of the G-E with Al_2O_3 using Carbide drill bit at (a) 15.08 m/min, 0.18 mm/rev and (b) 30.16 m/min, 1.40 mm/rev.

The SEM micrographs show the breakage of the fiber material and damage of the matrix material. Figure 11(a) shows the breakage of matrix material and also less damaged fibers but as the speed and feed are increasing the breakage of fibers as well as orientation of the fibers is seen with damaged matrix as seen in Figure 11(b).

CONCLUSIONS

From the experimental work, the drilling parameters, which are having influence on thrust and torque on the drilling of SiO_2/Al_2O_3 filled G-E composites, have been assessed.

- Unfilled and particulate filled G-E composites results reveal that thrust and torque depends on speed and feed. Further it is seen that with the increase in speed and feed, thrust and torque values showed an increasing trend.

- Alumina particulate filled G-E composite showed optimal thrust and torque values as compared to unfilled G-E and SiO_2 particulate filled G-E composites using carbide drill bit.

- SEM micrographs show that in drilling of G-E composites by carbide drill bit minimal material damage was observed. During drilling process the important SEM features observed were fiber fracture, cohesive resin fracture etc.
- From the experimental results carbide drill bit is the best tool for drilling of particulate filled G-E composites.

REFERENCES

1. J. Mathew, N. Ramakrishnan and N. K. Naik, "Investigations into the Effect of Geometry of a Trepanning Tool on Thrust and Torque during Drilling of GFRP Composites," Journal of Materials Processing Technology, Vol. 91, No. 1, 1999, pp. 1-11. doi:10.1016/S0924-0136(98)00416-6

2. I. El-Sonbaty, U. A. Khashaba and T. Machaly, "Factors Affecting the Machinability of GFR/Epoxy Composites," Composite Structures, Vol. 63, No. 3, 2004, pp. 329-338. doi:10.1016/S0263-8223(03)00181-8

3. R. Zitoune, V. Krishnaraj and F. Collombet, "Study of Drilling of Composite Material and Aluminium Stack," Composite Structures, Vol. 92, No. 5, 2008, pp. 1246- 1255.

4. E. Capello and V. Tagliaferri, "Drilling Damage of GFRP and Residual Mechanical Behavior—Part I: Drilling Damage Generation," Journal of Composites Technology and Research, Vol. 23, No. 2, 2001, pp. 122-130. doi:10.1520/CTR10920J

5. S. Arul, L. Vijayaraghavan, S. K. Malhotra and R. Krishnamurthy, "The Effect of Vibratory Drilling on Hole Quality in Polymeric Composites," International Journal of Machine Tools & Manufacture, Vol. 46, No. 3-4, 2006, pp. 252-259.doi:10.1016/j.ijmachtools.2005.05.023

6. H. Hocheng and C. C. Tsao, "Analysis of Delamination in Drilling Composite Materials Using Core Drill," Australian Journal of Mechanical Engineering, Vol. 1, No. 1, 2004, pp. 49-53.

7. S. C. Lin and I. K. Chen, "Drilling Carbon Fiber-Reinforced Composite Material at High Speed," Wear, Vol. 194, No. 1-2, 1999, pp. 156-162. doi:10.1016/0043-1648(95)06831-7

8. U. A. Khashaba, "Delamination in Drilling GFR—Thermoset Composites," Composite Structures, Vol. 63, No. 3-4, 2004, pp. 313-327. doi:10.1016/S0263-8223(03)00180-6

9. R. Piquet, B. Ferret, F. Lachaud and P. Swider, "Experimental Analysis of Drilling Damage in Thin Carbon/Epoxy Plate Using Special Drills," Composites Part A, Vol. 31, No. 10, 2000, pp. 1107-1115. doi:10.1016/S1359-835X(00)00069-5

10. V. Krishnaraj, S. Vijayarangan and A. Ramesh Kumar, "Effect of Drilling Parameters on Mechanical Strength in Drilling Glass Fiber Reinforced Plastic," International Journal of Computer Applications in Technology, Vol. 28, No. 1, 2007, pp. 87-93. doi:10.1504/IJCAT.2007.012336

11. A. M. Abrao, P. E. Faria, J. C. Campos Rubio, P. Reis and J. Paulo Davim, "Drilling of Fiber Reinforced Plastics: A Review," Journal of Materials Processing Technology, Vol. 186, No. 1-3, 2007, pp. 1-7. doi:10.1016/j.jmatprotec.2006.11.146

12. C. C. Tsao and H. Hocheng, "Evaluation of Thrust Force and Surface Roughness in Drilling Composite Material using Taguchi Analysis and Neural Network," Journal of Materials Processing Technology, Vol. 203, No. 1-3, 2008, pp. 342-348.doi:10.1016/j.jmatprotec.2006.04.126

13. V. Krishnaraj, S. Vijayarangan and G. Suresh, "An Investigation on High Speed Drilling of Glass Fiber Reinforced Plastic (GFRP)," Indian Journal of Engineering & Materials Sciences, Vol. 12, No. 3, 2005, pp. 189-195.

14. S. K. Malhotra, "Some Studies on Drilling of Fibrous Composites," Journal of Materials Processing Technology, Vol. 24, 1990, pp. 291-300. doi:10.1016/0924-0136(90)90190-6

15. J. Ramkumar, S. K. Malhotra and R. Krishnamurthy, "Studies on Drilling of Glass/Epoxy Laminates Using Coated High Speed Steel Drills," Journal of Materials and Manufacturing Processes, Vol. 17, No. 2, 2002, pp. 213-222. doi:10.1081/AMP-120003531

5

Feasibility Study of Boreholes Hand Drilling in Senegal —Identification Potentially Favorable Areas

Cheikh Hamidou Kane[1], Fabio Fussi[2], Moustapha Diène[3], and Déthie Sarr[1]

[1]Laboratory of Mechanics and Modeling-Sciences of Engineering, University of Thies, Thies, Senegal

[2]International Consultant, University of Milano Bicocca, Milano, Italie

[3]Department of Geology, FST-UCAD, University Cheikh Anta Diop of Dakar, Dakar, Senegal

ABSTRACT

Drilling techniques commonly used in Africa are rather well suited for areas where geologic formations are hard and groundwater is not located at higher depths. Thus, for a large number of people living in rural areas, access to improved drinking water sources is often limited, due to the high cost of drilled boreholes that is closely linked to geographical, geological and hydrogeological factors. The analysis of various contexts has revealed that, in order to improve access to safe drinking water for underserved communities and populations, it is possible to consider less costly alternative solutions, compared to current options for water supply which are still expensive. In this paper, a simplified drilling technology at a very low cost has been demonstrated: "the manual or hand drilling", which is a practical solution for less than 40-m deep water points in alluvial terrains or low resistance rock formations. The feasibility study of manual drilling in Senegal has revealed that, even if it is not practical in all geological formations of the country, manual drilling remains an alternative solution for reducing costs and improving accessibility to drinking water in several areas in Senegal, particularly in the Senegal River Valley, along the northern coast, in Fatick and Casamance coastal zones. This study was used to set up map of areas suitable for manual drilling boreholes; it aims to strengthen the local private sector capacity to meet growing drinking water needs in rural areas.

INTRODUCTION

Manual drilling borehole technique is a practical solution for less than 40 meters deep water points in alluvial soils and low resistance ground. Although it is not a practical solution in all geological formations, there are many areas in Africa where it can effectively provide drinking water at a very limited cost in rural areas. This is especially true in small isolated communities that will never benefit from major drinking water projects because they are not generally taken into account in national water policy. The objective of this paper is to identify suitable areas for manual drilling use in Senegal. This study was carried out in collaboration with UNICEF-Senegal, as part of a project to support the dissemination of manual drilling techniques in Africa. To set up this study, we first had to collect local information (at the major institutions in charge with water database) and then transfer it to UNICEF ftp site. These data were processed in collaboration with a group of UNICEF contracted international consultants in USA; they helped compile them in a geographic information system (GIS). GIS tools use makes it possible to analyze different thematic layers which have contributed to identifying, in each zone, the parameters that help define manual drilling feasibility.

CONTEXT

The Senegal Manually drilled boreholes are not commonly developed or popularized in Senegal. These drilling techniques have started to be developed in Francophone Africa in the late 80s and have never been promoted at wider scale. In Senegal, the technique of manual drilling is rudimentary and is used only in sandy areas (lithology without clay and hard rock). It involves a uniform-diameter hand-auger penetrating the ground at a depth not exceed 20 meters. The material used for casing is PVC and the screen slots are made manually and wrapped in a cloth (2 cm diameter slots in the PVC casing). In addition, the slotted party is not equipped with gravel pack, and wells are not usually developed. We also note that these boreholes are not often subjected to chemical analysis drilling acomplished. These manually drilled wells are often very common in the Niayes areas, in Casamance and exceptionally in Tambacounda (sedimentary ground at the alluvial zone, or in depression areas). In some terrains, although the water table is not deep, it is reached after penetrating clayey, lateritic, calcareous or marly layers. These shallow drilled wells (not exceeding 20 meters) tap water table aquifer and usually there are no lithostratigraphic data available after implementation, since these works most often carried out without permit. The unit cost of these wells ranges from 350,000 to 400,000 FCFA depending on the complexity of the ground and penetrated geologic formations; the investigation depths do not exceed in major cases 20 meters [1].

METHODOLOGY TO DEFINE SUITABLE AREAS

Criteria for Identifying Suitable Areas

Two main parameters were used to define each area potential level throughout the country and then to identify of favourable areas:

The Geological/Geomorphologic Ability

This involves identifying areas where subsurface layers show hardness and permeability characteristics suitable for manual drilled wells implementation; it is about the possibility to manually perform a small diameter shallow borehole in permeable ground that could yield a significant flow rate.

The Ability Based on Groundwater Depth

This consists of identifying areas where you can find exploitable flowing groundwater at a depth compatible with manual drilling techniques. In Senegal, it might not be useful to set up a morphological classification based on the elevation model, to identify morphologically favourable areas; the following reasons can be evocated:

- The ground morphology is almost flat and then automatic process creating different morphological areas does not give acceptable results;
- The occurrence of favourable superficial deposits (not mentioned in the geological maps) is already taken into account in available morpho-pedological map.

In addition, areas located in Eastern Senegal, where relatively high differences in altitudes are occurring, are considered in our analysis. We do consider as well the presence of weathered layers that thickness varies locally with respect to topography. These aspects are not well documented in morphologic and soil maps.

Determination of the Geological Suitability

This mainly consists of identifying areas with subsurface layers characterized by hardness and permeability suitable for hand drilled boreholes. These terrains are made of primary geological layers or weathered/sedimentary formations underlain by the main geological formation.

Information Used

Main information sources used in this study are listed below [2]:

- Senegal aquifers map, published by the Water Management and Planning Department in 2008, available in soft copy (Figure 1);
- Morphological and soil map (Figure 2);
- Water points database (dug wells and boreholes);
- Stratigraphic data of boreholes logs.

Upon the moment we completed the study, the electronic version of Senegal geological map was not available. Consequently we have used aquifers map to characterize general geologic features, in addition we integrate information provided by morphologic and soils maps.

Definition of the Geographical Unit to be Analysed

Based on observation and crossed-cutting analysis of different information levels, the study found that aquifers units mentioned on the map are generally overlain by thick geologic layers generated by sedimentation or weathering process. Thus, hardness and permeability characteristics of the main aquifer recorded on the map, do not match in most situations with the lithology currently existing in the first 30 - 40 meters (interesting fringe for manually drilled borehole). We do notice that on the same aquifer unit, subsurface layers features are very different.

At the other hand units recorded on the morphological and soil map are more in line with the current subsurface layers features; however they may show various characteristics depending on geological formations they belong to.

To identify potential areas that are characterized by the same type of subsurface layers, we have used a combination of aquifer units (Figure 1) and morphopedological units (Figure 2) as a criterion for defining geometric basis for the geological suitability classification. The two combined units are thus considered as two information layers (aquifers and morphopedology), which produced a vector layer where each polygon is characterized by a specific combination of morpho-pedological units and aquifer.

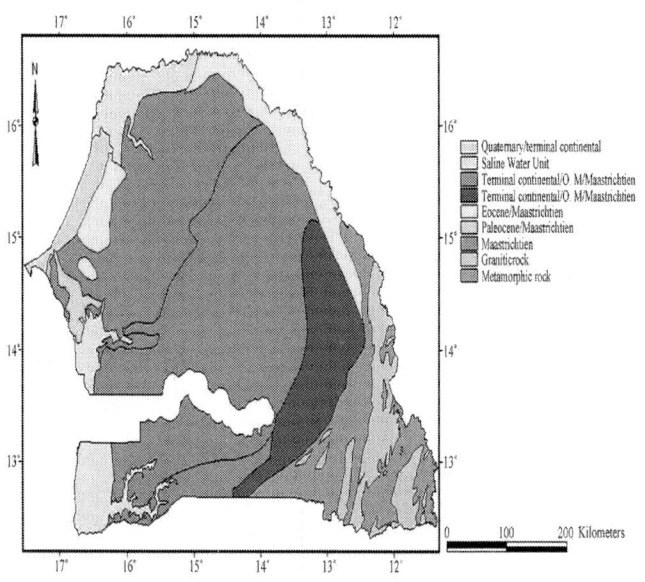

Figure 1: Senegal aquifers map (source: Water Resources Management and Planning Department).

Morpho-pedology units

Sandy formations: string and terrace sand Sandy formations: littoral dunes
Allomorphic soils
Hydromorphic soils
Ferralitic soils
Ferruginous soils on erg
Ferruginous soils on deep sediments
Ferruginous soils on sediments: littoral, less develop soils
Ferruginous soils on deep primary materiau
Ferruginous soils

0   200   400Kilometers

Figure 2: Morphologic and soils map.

Assessment of Subsurface Layers Characteristics

For each unit (morphopedological +aquifer) we have considered:

- Lithology and texture of the most superficial layers derived from boreholes logs (considering only layers which thickness exceeds 5 meters);
- The thickness and depth of all lithological units, lightly or meanly consolidated (hardness);
- The spatial occurrence of boreholes penetrating lateritic levels and their percentage (Figure 3);
- The presence of manual wells, and their relationship with existing boreholes (Figure 4).

Drilling log data (Figure 3) were used to assess depth of low resistance geologic layers with regard to their hardness characteristic and laterite occurrence as well. Data processing has showed that almost all borehole logs throughout the country encounter low resistant geologic layers, within the first 50 meters, which might be penetrated manually. In other words, there are favorable conditions for manual drilling, in terms of subsurface layers hardness. At the same time it is obvious that, at the central and southern parts of the country interbedded lateritic layers with 3 - 5 meters thickness are commonly found in borehole logs. In other words, if the lateritic layers can not be considered as an obstacle to manual drilling, it is more adequate to adopt (in major country parts) appropriate techniques suitable for penetrating through a few meters lateritic layers.

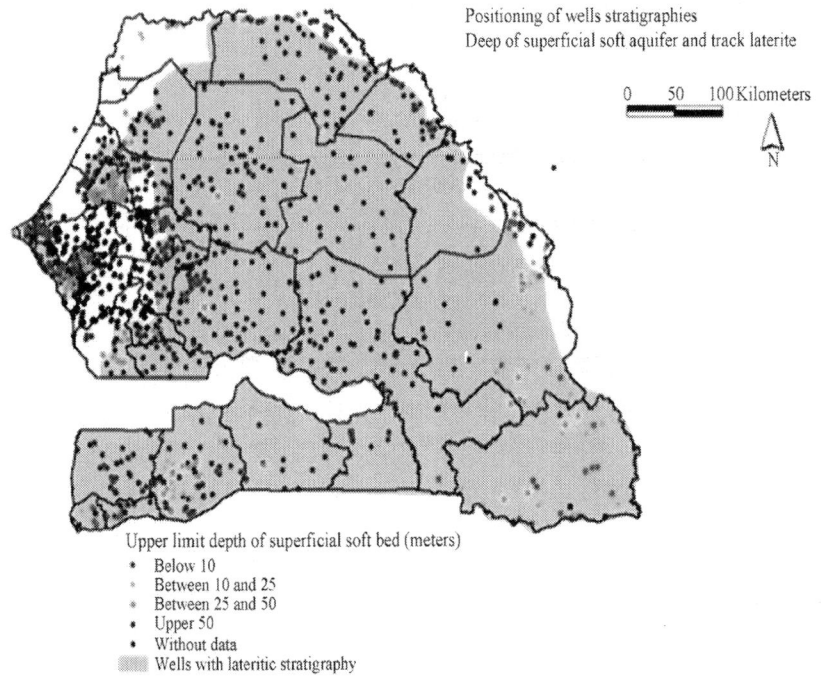

Positioning of wells stratigraphies
Deep of superficial soft aquifer and track laterite

0   50   100 Kilometers

N

Upper limit depth of superficial soft bed (meters)
• Below 10
  Between 10 and 25
• Between 25 and 50
• Upper 50
• Without data
▨ Wells with lateritic stratigraphy

Figure 3: Subsurface layers characteristics derived from available bore-holes logs.

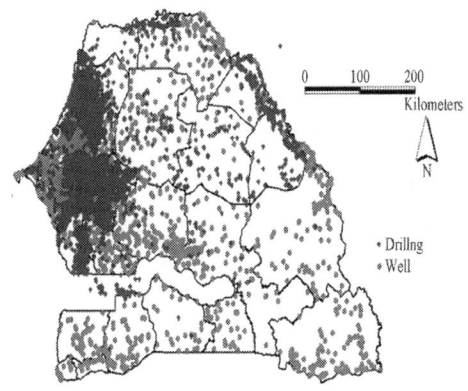

Figure 4: Dug wells and boreholes distribution map.

From the observations results we came out to an assessment of subsurface layers characteristics (such as average features up to 30 meters depth) for each aquifer unit (morphopedological +aquifer). The following aspects have been defined:

- A description of probable lithology subsurface;
- Estimated subsurface layers hardness;
- The subsurface layers permeability;
- The geological suitability class.

Allocation of Geological Suitability Classes

Based on subsurface layers characteristics, a geological suitability value was assigned to each area. The following classes are considered in geological suitability classification (Figure 5):

- FO: High geological aptitude area that corresponds to very favorable conditions of hardness and permeability on the main aquifer formation;
- FO-alt: High geological aptitude area of weathered layer composing the main aquifer formation;
- FOR-dep: high geological aptitude area of sedimentation deposits composing the main aquifer (deposits are generally consistent with the morpho-pedological description);

- MO: average geological aptitude of layers with hardness features that correspond to moderately suitable aquifer rock;
- MO-alt: average geological aptitude area of sedimentation deposits composing the main aquifer (deposits are generally consistent with the morpho-pedological description);
- FA: Poor geological aptitude area, corresponding to hydrogeological environment that is generally not suitable for manual drilling.

RESULTS ANALYSIS METHODOLOGY

Suitability Determination Based on Geological Features

We can observe that the country's major part shows geological environments with average aptitude for manual drilling. Geologically, the Senegal River Valley area [3], the Niayes zone of Northern Senegal [4], and coastal area of Southern Senegal are considered as very suitable.

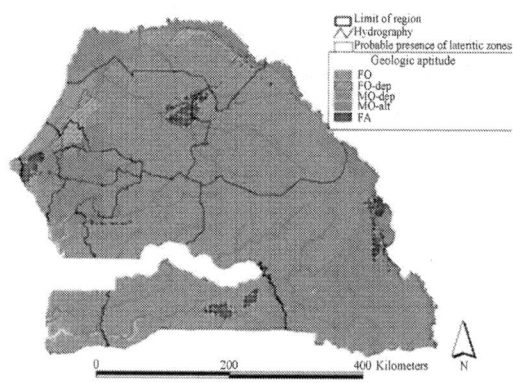

Figure 5: Map of geological aptitude to manual drilling.

Other less important areas show high geological limitations and are considered as areas not favorable for manual drilling.

In the South-East country part, the geologic aptitude is considered as average due to weathered formations occurrence; in that area, we have crystalline basement, hard consolidated rocks that may become partially favorable with respect to weathered shallow layers occurrence. Therefore, the feasibility of manually drilled wells will depend either on weathered deposits accumulation related to depression (or low slope), or on rocks permeability (crystalline rocks, especially the fine-textured rocks produce weathered layers with clay presence). It should be noted that throughout the central area there are lateritic layers. If those layers are usually perforated with manual techniques (when thickness is thin) it must be accepted that percussion techniques are the most appropriate ones.

Suitability Determination Based on Water Table Depth

In Senegal, there is database containing information over 7000 recorded water points (wells and boreholes), particularly in the western and northern area (Figure 6). These data can provide relevant information to assess water table depth. However, in the central and the eastern part of the country, the quality of information is poor. In order to assess manual drilling suitability with respect to water table depth, we have considered directly average water table depth provided by the available groundwater depth map.

The following aspects have been taken into account:

- The water points' density is very different from one area to another, and in some of them the distance between recorded water points is long. Therefore, in order to get reliable results, the interpolation algorithms require reference data' distribution to be as homogeneous as possible and compatible with the size of interpolation mesh;

The local water depth variations depends on groundwater piezometric level' shape, as well as local topographic variations, which values are not recorded in water points database (Records consider water table depth, but not absolute static level in terms of elevation above sea level).

The following categories/classes were used to estimate suitability based on water table depth (Figure 7):

- Water table depth under 10 m: compatible with different manual drilling techniques, suitable;
- Water table depth between 10 to 25 m: manual drilling is possible. For this category all the drilling techniques are applicable, however manual drilling will require more consideration to resistant layers occurrence that may prevent to reach expected depths;
- Water table depth beyond 25 m: generally manual drilling techniques do not match with these depths.

We can notice hat along the coast, in Casamance area as well as the eastern part of the country, manual drilling is possible if we refer to water table depth. On the other hand, manual drilling techniques use is very difficult, in the central part of the country, because groundwater depths are generally high. The information available for establishing water table map is insufficient. Consequently, the assessment was carried out with limited water points' number, which shows out huge differences in water static levels. In this context, the general trend of water level distribution does not clearly emerge. Therefore, more detailed analysis requires complementary information along with field observations that provide accurate indication on shallow groundwater depth in targeted areas.

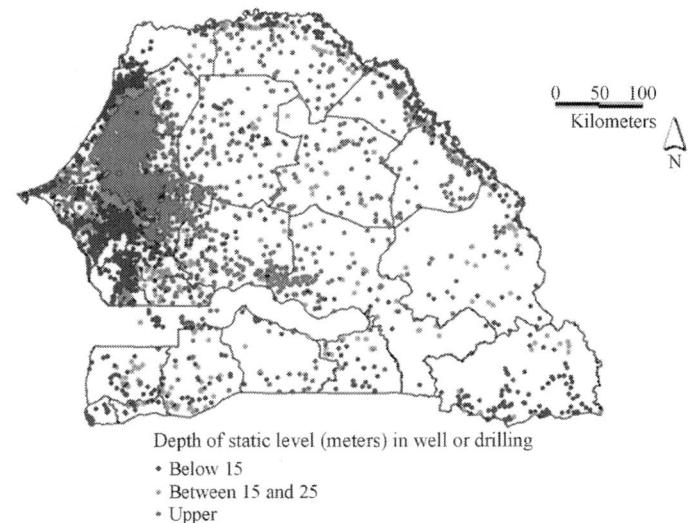

Depth of static level (meters) in well or drilling
• Below 15
• Between 15 and 25
• Upper

Figure 6: Map of water table depth.

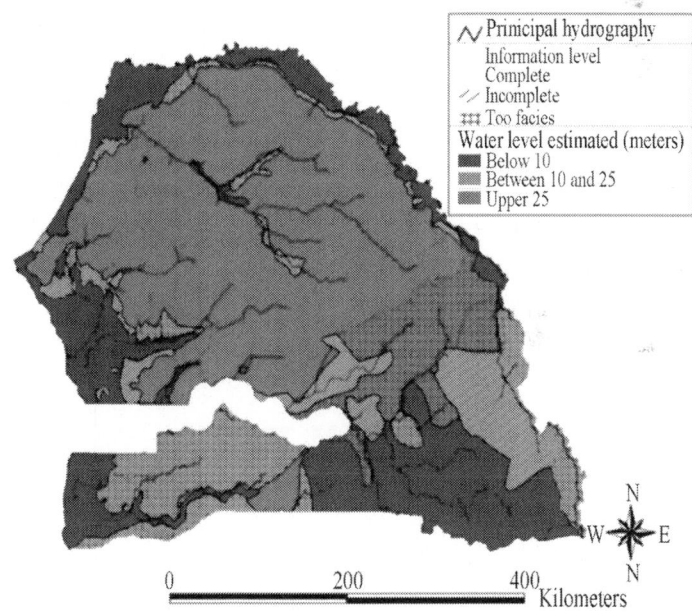

Figure 7: Map of assessed water table depth.

Total Suitability Determination

The final assessment of suitability for manual drilling techniques in Senegal was done based on cross-cutting information provided both by geological suitability and aptitude related to water table depth.

The following suitability classes were defined with regard to geological features and groundwater depth as well (Figure 8):

- Very favorable: refer to areas where geological formations features and water table levels are both favorable;

- Favorable: refer to areas where, one parameter shows an average aptitude for manual drilling use, and on the other is rather favorable;

- Little-favorable: refer to areas where both parameters show an average aptitude (i.e. favorable with some limitations); in these areas manual drilling techniques may be used depending on topographic conditions, however in general these areas have constraints for implementing these techniques. For instance in the eastern part of the country, undulating topography (with low altitude zones) and weathered layers occurrence may provide specific environment for manually drilling techniques' use.

- Not favorable: refer to areas where, either one or both parameters have suggested unfavorable conditions for manual drilling; therefore these techniques are generally difficult to use.

RESULTS

The Country Central Area

This area is considered not suitable for manual drilling mainly because of water table level, which is relatively deep [5]. At the western, southern and northern parts (Senegal River valley) encounter very favorable areas for manual drilling techniques' use; these are listed below:

The Valley Area and Along Senegal River

This zone consists of a 10 - 25 km strip along the Senegal River starting from Bakel to the River mouth. It is part of administrative regions of Saint-Louis, Matam and Tambacounda. That eco-geographical area lays between two climatic zones: the Sahelian from Northwest to Northeast, and Sudanian domain in southeast. In the Sahelian zone the mean annual rainfall is 200 - 300 mm; this area is subject to pluviometric deficit and irregular rainy season, irregularity with relevant interannual variations. In the Sudanian zone rainfall is about 500 mm/year.

Based on the information from morpho-pedological map and boreholes logs as well, the subsurface layers are supposed to be made of unconsolidated sediments characterized by various permeability (clay, sandy clay, or alternatively sand). The geological formations' features show that it is possible to set up manual drilling in these areas; however, low production yield may be the main constraint.

The Northern Coastal Area

It extends along the coastal strip north of Senegal and covers the administrative regions of Dakar, Thies, Louga and St Louis. The area is characterized by Quaternary sedimentary formations that overlay older Secondary and Tertiary geological formations (Maastrichtian, Paleocene and Eocene). These Quaternary layers extend to major country part and are made of sandy material.

The available and accessible water resources supplying people including farmers are mainly from two sources:

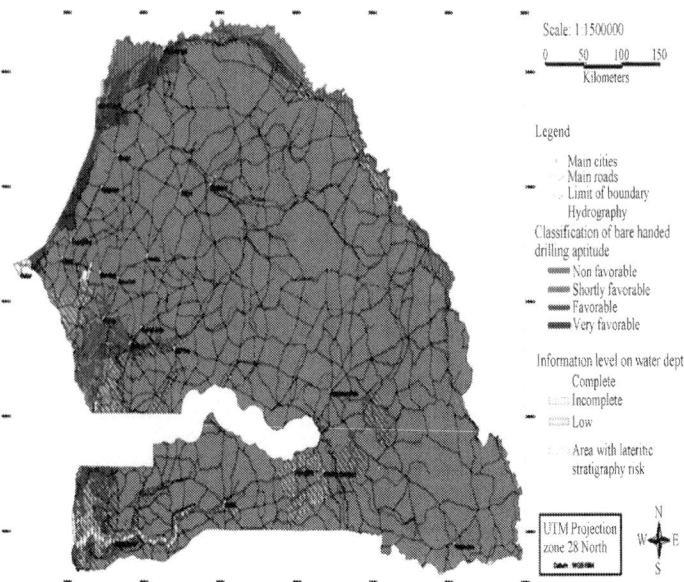

Figure 8: Map of areas suitable for manual drilling.

groundwater and surface water; the latter consists of local lakes and ponds at reduced number, which usually constitute main outlets of rivers system active only in rainy season [4,6].

If we refer to geological features and water table depth, this area is considered to be the most favorable one to manual drilling. It is necessary to pay attention, locally in some areas, to saline intrusion risk. Consequently, it is appropriate in this case to develop a water supply strategy using hand pump technologies. This strategy is suitable for small rural communities, which water demand is limited; by this way saline intrusion risk can reduce in this coastal aquifer. The resulting yield of manually drilled water points can easily meet water supply demand in these areas.

The Coastal Zone in Fatick Region

In this area, groundwater is not very deep and occurring subsurface layers are made of sandy clay and sand sediments. It is considered as favorable zone, with limited borehole yield, particularly in ground bearing clay in the consolidated deposits.

Casamance Zone

In this area, the geological conditions are generally favorable, they consist of sandy clay or clay layers. From the perspective of groundwater level, shallow water depth areas are identified in the western part (Ziguinchor administrative region), whereas areas with relatively deep water table are located in the central part (Kolda region).

That delimitation implies considering Casamance west part as favorable to hand drilled wells. In reverse its central part is partially favorable, since drilling depth may be beyond 20 m, and in addition interbedded more consolidated layers may be found there. In this context, more appropriate drilling techniques must be used (lateritic layers).

CONCLUSIONS

The design and construction of hand drilled wells using appears to be a practical solution for water points tapping shallow groundwater (40 m deep) occurring in alluvial soils or slightly consolidated formations. Even though it is not a suitable solution in all geological formations, this study has identified favorable areas in Senegal where this technique can be applied to provide drinking water for rural populations at a very limited cost. In Senegal (Casamance), a manually drilled well fitted with hand-pump costs approximately USD 3300, including bore-hole/ well, pump, superstructure and supervision. At the other hand, percussion drilled borehole with the same equipment costs USD 14,300 in Kedougou, and a 30-m deep dug well fitted with the same surface equipment costs more than USD 16,000.

REFERENCES

1.   C. H. Kane, "Etude sur l'optimisation du coût des Forages en Afrique de l'Ouest—Rapport Sénégal, Financement Banque Mondiale," 2007.

2.   PAGIRE, "Plan d'Action pour une Gestion Intégrée des Ressources en Eau, Rapport Sénégal, DGPRE," 2007.

3.   M. Audibert, "Delta du Fleuve Sénégal Etude Hydrogéologique Projet Hydroagricole du Bassin du Fleuve Séné- gal," 1970.

4.   SENAGROSOL/EDE, "Etude d'impact environnemental et sociale des travaux de mobilisation des ressources alternatives pour l'irrigation des Niayes," 2009.

5.   PSE/COWI/POLYCONSULT, "Etude Hydrogéologique de la Nappe Profonde du Maastrichtien."

6.   SENAGROSOL, "Evaluation Environnementale Stratégique (EES) des activités relatives à la promotion de la micro irrigation dans les Niayes, le Bassin Arachidier élargi a la région de Tambacounda et à la Casamance," 2009.

6

Potential Implementation of Underbalanced Drilling Technique in Egyptian Oil Fields

K.A. Fattah[a], S.M. El-Katatney[b], and A.A. Dahab[b]

[a]Petroleum and Natural Gas Engineering Department, College of Engineering, King Saud University, Riyadh 11421, Saudi Arabia
[b]Petroleum Engineering Department, Faculty of Engineering, Cairo University, Egypt

ABSTRACT

The need to increase productivity and to reduce drilling damage favors the use of underbalanced drilling (UBD) technology. In highly depleted reservoirs, extremely low-density fluids, such as foams or aerated mud, are used to achieve circulating densities lower than the pore pressure. In such cases, the induced modification of the in situ stresses has to be supported mainly by the rock, with little contribution from the drilling fluid pressure. The application of underbalanced drilling depends on the mechanical stability of the drilled formation, among other factors. In general, poorly consolidated, depleted formations are not suited for that technology.

In this paper, 23 UBD worldwide cases have been analyzed; two of which are from Egyptian fields and the others are from Iran, Algeria, Kuwait, Oman, Texas, Mexico, Indonesia, Canada, Libya, Middle East, Qatar, Saudi Arabia and Lithuania. From these analyses, the reasons of failure or success have been stated. The reasons of success included depleted reservoirs and highly fractured carbonates formation while, the reasons of failure include over pressurized shale, highly tectonic stress areas, and downhole failures. The main attractive application of this technology was proposed to be only in the reservoir section, and the target was to prevent the reservoir damage and hence increase the productivity and recovery factor.

A proposed underbalanced drilling program is developed based on these analyses to be used in the three main regions in oil and gas producing Egyptian fields. The aerated mud was selected as a drilling fluid to drill the reservoir section in Western Desert and Gulf of Suez region whereas the single phase fluid was selected as a drilling fluid in the Nile Delta region.

INTRODUCTION

Drilling cost is considered one of the major components of operating cost in the petroleum industry. Improving the penetration rate of drilling operation and reducing drilling problems, such as pressure differential pipe sticking and lost circulation, have long been considered an effective way of decreasing drilling costs. The overbalance pressure, generally recognized as the most important among the many factors affecting penetration rate, is often defined as the pressure differential between the borehole pressure and formation fluid pressure (Murray and Cunningham,

1955, Eckel, 1957,Cunningham and Eenink, 1959, Gamier and van Lingen, 1959, Vidrine and Benit, 1968,Bourgoyne and Young, 1974a, Bourgoyne and Young, 1974b and Black and Green, 1978). Formation pressures lower than the static pressure of a column of fresh water require the use of a lighter fluid, such as air, injected with liquid to obtain lower overbalance pressure to enhance penetration rate and to minimize lost circulation and pipe sticking as well as formation damage. Therefore, aerated mud drilling "implies the use of air or natural gas as the circulating medium instead of the regular mud" is becoming an attractive practice in some areas. The commercial use of aerated mud drilling began only in recent years (Rankin et al., 1989 and Claytor et al., 1991). Low-density drilling fluids used in underbalanced drilling consist of air, mist, stable foam, and aerated mud foam with back pressure. Whereas the term "aerated mud" implies the simultaneous introduction of air and mud together into the standpipe in order to drill special types of formations (Godwin et al., 1986, Boyun and Rajtar, 1995 and Salah El-Din and El-Katatney, 2009).

The main advantage of air as a circulating fluid is that being the lowest density fluid. It imposes minimum pressure on the formation to be drilled. High penetration rates have been achieved in hard and dry formations with the use of air as a circulating fluid. In addition to high penetration rate, longer bit life results through the use of this medium as compared to mud. Drilling rates as high as 90 ft/h have been attained in shales. Air drilling, however, is restricted to areas where high volume water sands are not present ahead of the producing zone. The rate of water influx that can be handled in the case of air drilling is also not well known. Other inherent disadvantages of using air or natural gas as drilling fluids include possibility of downhole fires and explosions, and sloughing of formations due to underbalance of stresses around the wellbore. Possibility of downhole explosions are of particular concern in air drilling operations. Small dust-like particles are generated as a result of rock cuttings (chips) being ground and pulverized by the drill string in the annulus, and collision of cuttings with each other, the tool joints, and the wall of the borehole due to the high velocity forces. In the presence of moisture, seal rings may form at tight places in the annulus, which create pressure chambers. With additional influx of natural gas from gas-bearing zones being penetrated by the bit, an explosion may easily occur.

Besides having formations suitable for air drilling, the most important consideration in drilling with air is the volume of air required. Air drilling often fails because of insufficient volume of air necessary to clean the hole efficiently under certain conditions, e.g., wet hole, sloughing shales, and influx of formation water. A practical rule of thumb for determining adequate air volume is that the volume required achieving 1000 ft per minute annular velocity to clean the hole properly (Godwin et al., 1986 and Boyun and Rajtar, 1995).

Drilling with foam has some appeal due to the fact that foam has some attractive qualities and properties with respect to the very low hydrostatic densities, which can be generated with foam systems (Hooshmandkoochi et al., 2007, Moore and Lafave, 1956, Maurer, 1998 and Bentsen and Veny, 1976). Foam has good rheology and excellent cutting transport properties. The fact that foam has some natural inherent viscosity as well as fluid loss control properties, which may inhibit fluid losses, makes foam a very attractive drilling medium. During connections and trips, the foam remains stable and provides a more stable bottom hole pressure. It is a particularly good drilling fluid with a high carrying capacity and a low density. The foam normally remains stable, even when it returns to the surface, and this can cause problems on a rig if the foam cannot be broken down fast enough. In earlier foam systems, the amount of defoamer had to be tested carefully so that the foam was broken down before any fluid entered the separators. In closed circulation drilling systems, stable foam could cause particular problems with carry over. The recently developed stable foam systems are simpler to break, and the liquid can also be refoamed so that less foaming agent is required and a closed circulation system can be used. These systems, in general, rely on either a chemical method of breaking and making the foam, or the utilization of an increase and decrease of pH to make and break the foam. The foam quality at surface used for drilling is normally between 80% and 95%. The quality of foam means that the system is 80–95% gas, with the remaining 5–20% being liquid. Downhole, due to the hydrostatic pressure of the annular column, this ratio changes as the volume of gas is reduced. An average acceptable bottom-hole foam quality (FQ) is in the region of 50–60%. Fluid densities for foam range from 1.6 ppg to 6.95 ppg (0.2–0.8 S.G.) (Godwin et al., 1986 and Boyun and Rajtar, 1995). The density ranges are adjusted with the make up of the foam by adjusting the Liquid Volume Fraction (LVF) through the injection of liquid and gas by adjusting the backpressure on the well. The backpressure adjusts the downhole pressure and slows down the velocities in the annulus. Experience has proven that foam is able to handle over 100 bbl/h of water influx (Godwin et al., 1986, George and Waston, 1956 and Boyun and Rajtar, 1995).

So, the objective of this research work is to investigate and analyze many worldwide applications of underbalanced drilling and state the reasons of success or failure of this application. Based on these analyses, a proposed underbalanced drilling program is developed. In this proposed program, the method of selecting the appropriate technique to be applied for these candidate areas are selected according to the geology of the area and the bottom hole conditions inside the wells.

STUDIED CASES

In this section, three case studies from Egyptian fields and other places are analyzed in detail and a summary of 20 cases from other worldwide fields are given with a brief discussion about their objectives, problems and results (Salah El-Din and El-Katatney, 2009).

Case 1: Gulf of Suez Area

The well is located at onshore Belayim oil field. The well target was sandstone of zone III (Belayim formation, Feiran member) at a total depth of 2335 m TVD, 2854 m MD. The pressure in Zone III (sandstone) was estimated to be 3000–3500 psi (0.3917–0.4569 psi/ft). The objectives of UBD were to increase rate of penetration, enhance Well control, reduce occurrence of lost time incidents, and increase well productivity. The 20 m of the new hole at 7 in. liner shoe at 2659 m MD was drilled with only mud, then the MWD signal test was performed (inflow test and also to test the optimum rate combination for better MWD signal) as shown in Table 1. Based on this test, the formation pressure was estimated to be less than 2500 psi that was confirmed at 2400 psi from vacuum test and the MWD can work up to 21% nitrogen. Nitrified mud (500 SCFM + 230 gpm diesel) was applied while close balance drilling the 6 in. original and side-track lateral section. The 6 in. hole was drilled to depth 2830 m utilizing UBDS and powerpack motor of 1.15° BH c/w MWD Impulse, VPWD, ADN tools (inclination at bit, annulus and string pressure, GR resistivity, density-neutron) with 2 × 3-1/2 in. W.FORD float valve + motor restriction sub (nozzle 14/32 in.) for improving MWD signal. The analysis of this well results showed that, The ROP was enhanced drastically in sand from 4 m/h while sliding to 50 m/h, and in anhydrite was 8–10 m/h (experienced 2–4 m/h in normal overbalance drilling), the use of rotating head helped to control well while tripping and also in case of separator carry over problems, and the Crew acquired UBD work experience.

Table 1: Change in BHCP versus mud rate and N2 rate

Dura-tion (h)	Mud rate (GPM)	N_2 rate (SCF/m)	N_2 (%)	SPP (psi)	BHCP (psi)	ECD (kg/lit)	Gain (bbls)	MWD Signal
1.5	250	250	4.8	2250	2887	0.86	31 mud	Ok
3.0	250	500	9.1	2250	2660	0.79	34 mud	Ok
2.0	240	500	11.4	1800	2652	0.79	0	Ok
1.5	230	500	13.2	1600	2620	0.78	0	Ok
1.5	210	500	15.5	1450	2590	0.77	0	Ok
1.5	180	500	21.6	1120	2549	0.76	0	Ok

Case 2: Western Desert Gas Field Area

The well is located at the central part of the western desert block. The well target was to drill 3-7/8 in. × 500 m horizontal section in unit 3 of the Mesozoic Lower Safa reservoir. They are composed of low to medium permeable (1–500 md) micaceous sandstones deposited in a strong tidally influenced estuary, Fig. 1. Lower Safa formation comprises a high-energy sequence of Estuarine deposits with a total average thickness of 110 m in the area where is planned, although only 29 m of these thickness are considered productive. The objective of UBD was to prevent reservoir damage. Gasification was through drill pipe injection technique.

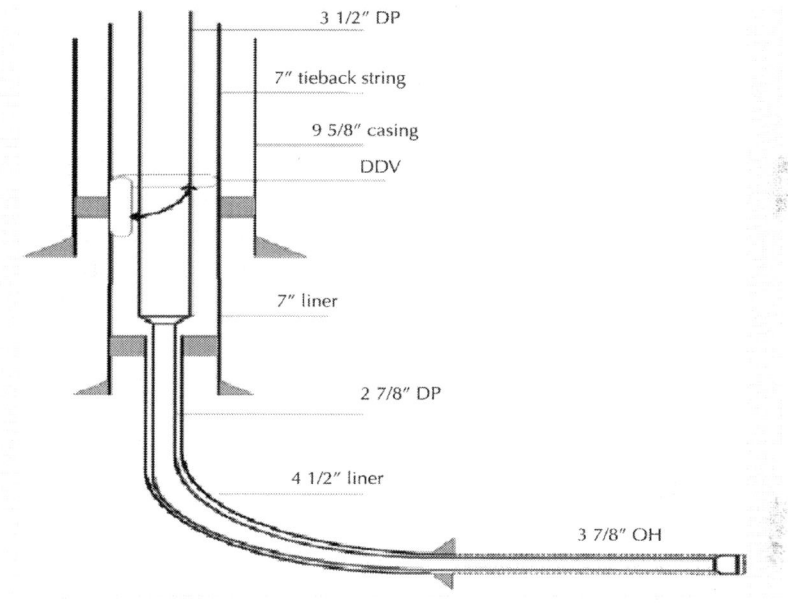

Figure 1: Well profile diagram for case 2: western desert gas field area.

The well was completed as open hole. Average ROP during overbalanced drilling operations on offset wells has been historically 2–3 m/h in the horizontal section. Historical data for UBD wells suggested that there will be an improvement in ROP due to the elimination of the chip hold-down effect. It was estimated that the ROP will be between 5 and 10 m/h. The drilling fluid of choice was produced water. The drilling fluid could be separated from the produced hydrocarbons and re-used. Due to the CO_2 content of the reservoir (up to 9%) and the use of nitrogen (up to 5% O2), corrosion mitigation was required. Once the well started to produce during the drilling phase, the N_2 was stopped, which in turns eliminated excessive use of corrosion inhibitors. Water and nitrogen gave the desired underbalanced margin when kicking off the well, and water was treated with suitable chemicals for corrosion mitigation. It became apparent that the Lower Safa formation was normally pressured. Hence by using just water, the BHP will be 260 psi underbalanced. Nitrogen was required to create a greater draw down than the 260 psi as it is unknown at what draw down the matrix starts to contribute to the inflow.

As soon as the well produced, nitrogen was cut down to zero rates. Nitrogen injection was required again every time the drill string tripped through the Down-hole Deployment Valve (DDV) to remove the water from the reservoir section.

Fig. 2 shows the working window (operating envelope) for the well (case 2) with no reservoir inflow for, 3-7/8 in. hole, 3-1/2 in. × 2-7/8 in. drill pipe design, 2 × 500 m legs, and bit at TD. Also plotted on the operating envelope, are the various constraints that must be fulfilled during underbalanced drilling operations. After drilling 200 m, the drilling had been stopped due to failure of downhole equipment due to high temperature.

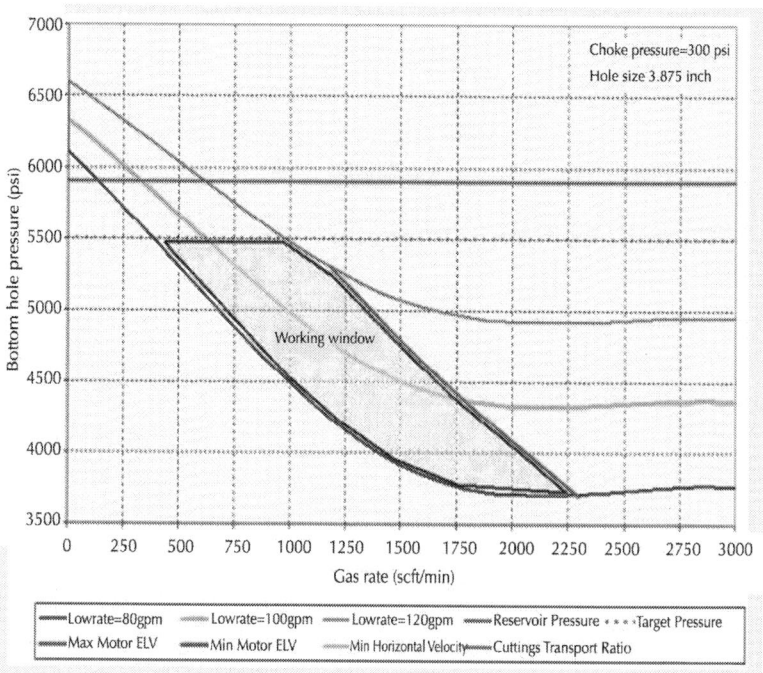

Figure 2: Working window for case 2: western desert gas field area.

Case 3: Iranian Oil Field

The target reservoir for this well was Asmari formation, the formation was fractured carbonated formation. The reservoir drive mechanism was gas cap. Shale strings were not expected in this formation. Expected reservoir pressure and temperature were 2622 psi and 141 °F, respectively. Reservoir fluid was oil with API gravity of 25°, GOR 564 SCF/STB, and H2S concentration of 240 ppm. The permeability of the reservoir was 0.1–1000 md with a porosity of 9% (Hooshmandkoochi et al., 2007). The well was drilled from m (9-5/8 in shoe depth) to a total depth of 2938 m MD (2567 m TVD), Fig. 3. The primary objectives of this underbalanced drilling project were to: minimize drilling induced formation damage, eliminate drilling fluid losses, and improve drilling performance. The drilling fluid selection was one of the most critical decisions in planning an underbalanced well. The right fluid(s) selection will not only lead to suitable BHCP but will also minimize pressure transients and thus eliminating/minimizing formation impairment. The deviated underbalanced section of this well was to be drilled with a Gachsaran field native crude oil and a membrane nitrogen generation circulating system. Liquid Phase, the native crude oil, was chosen over Diesel and other drilling fluids because it is the natural reservoir fluid for this well. This minimized chances of formation damage in event of pressure transients and/or from fluid imbibitions. The well was displaced with the produced fluid after getting enough oil production. Gas Phase, nitrogen, was selected as the injection gas because of its inert nature, economic availability and suitability for this specific underbalanced drilling project. Nitrogen was obtained from the surrounding air and generated onsite, by nitrogen production unit (NIOC's). The multiphase flow behavior in the wellbore during underbalanced drilling was very complex. The response of the downhole conditions to changes in various flow parameters must be characterized prior to the commencement of underbalanced drilling operations in order to maximize chances of success. Fig. 4 contains a plot of the bottom hole circulating pressures induced by a variety of nitrogen rates and the Gachsaran native crude oil injection rates. This plot was referred to as the operating envelope. Also plotted on the operating envelope, are the various constraints that must be fulfilled during underbalanced drilling operations. The range of flow rates that satisfy all of the constraints, defined the acceptable operating region. A minimum drawdown at the bit of 200 psi was required to ensure adequate underbalanced conditions in the well, with a maximum drawdown of 300 psi to minimize any near wellbore depletion effects. The target bottom hole circulating pressure at the bit for this well was 2300–2400 psi.

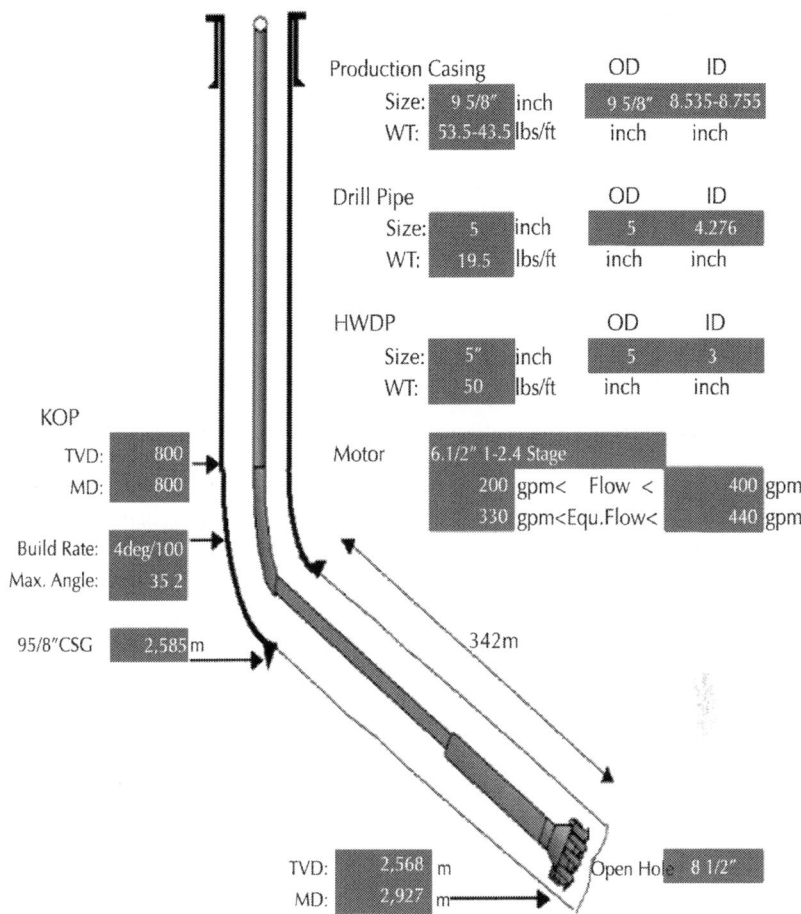

Figure 3: Well profile diagram for case 3: Iranian oil field area.

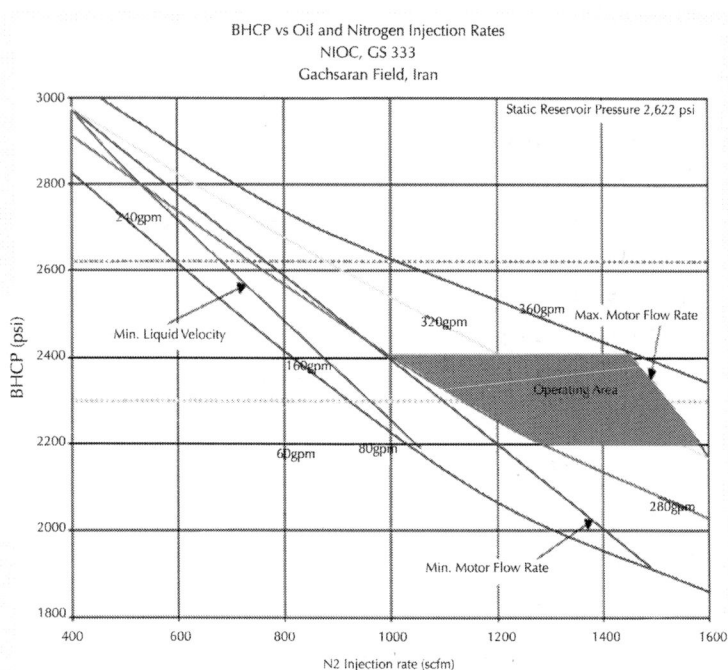

Figure 4: Operational envelope – native crude for case 3: Iranian oil field area.

UBD on this well experienced some typical logistical and start up problems associated with a steep learning curve, this being the first such operation in Iran. Despite all the problems encountered in this well, the following performance had been achieved: drilled to 308 m of total open hole depth, no loss circulation was encountered while drilling, successfully implemented UBD technology, and no Quality, health, safety and environment (QHSE) incidents were recorded Bennion et al., 1998, Dorenbos and Ranalho, 2002, Gordon, 2005, Gray, 1957, Hongren et al., 1999, International Association of Drilling Contractor, 2005, Kuru, 1999, Louison et al., 1984, Maclovio, 1996, Meng, 2005, Moore et al., 2004, Nas, 2004, Negra et al., 1999, Parra et al., 2003,Qutob, 2007, Qutob and Ferreira, 2005, Sunthankar, 2001, Weatherford Company, 2006, Westermark, 1986, Whiteley and England, 1986 and Zhou, 2005).

Data Analysis

The following analysis is carried out based on some actual wells drilled underbalanced worldwide. As mentioned before, the main advantage of underbalanced drilling techniques is to increase the rate of penetration as compared with overbalanced drilling techniques.

Table 2 gives the recorded data that were collected from successful underbalanced drilling cases in which the aerated mud was used to drill sandstone reservoir sections (Moore and Lafave, 1956).

Table 2: Recorded ROP in Algeria

Algeria sandstone reservoir		
Well number	ROP overbalanced (ft/h)	ROP underbalanced (ft/h)
1	10.4	19.5
2	10.4	17.6
3	19.3	22.5
4	19.5	22.3
5	13.5	45
6	17	26.6

From Fig. 5, there is an observed increase in ROP in all cases that were drilled by underbalanced techniques. In underbalanced drilling, ROP was increased due to the disappearance of chip hold-down effect. So the normal trend includes that an increase of the ROP resulted from a decrease in the hydrostatic pressure of drilling fluid as compared with the pressure of the formation that drilled by UB, as shown in Fig. 6.

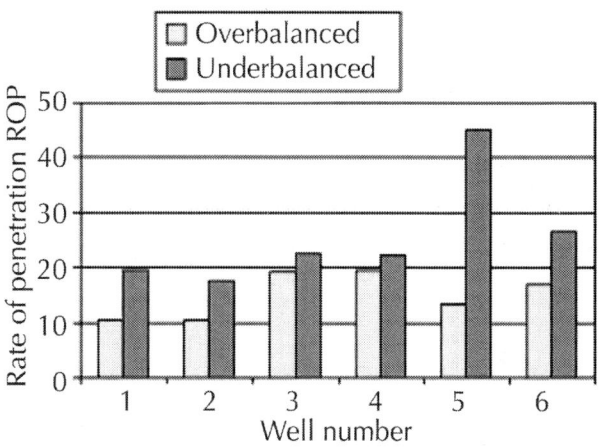

Figure 5: Comparison between ROP in OB and UB cases.

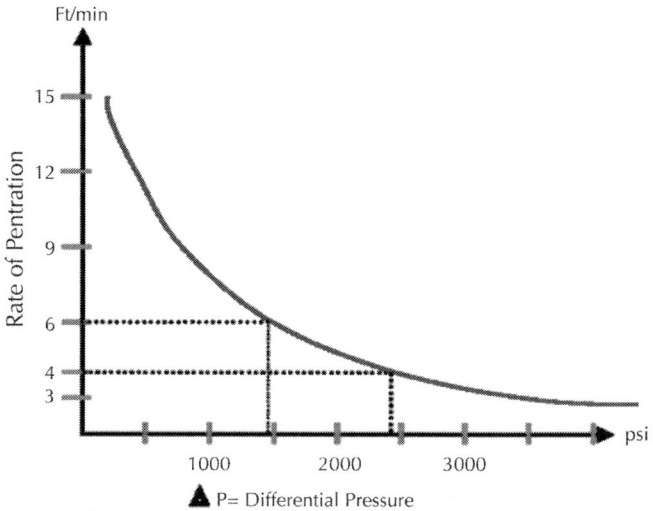

Figure 6: Relationship between ROP and pressure drop.

Table 3 gives the recorded data of ROP (ft/h) and pressure drop (psi) for different reservoirs that were drilled by aerated fluid as an UBD drilling fluid. These reservoirs have the same lithology but having different reservoir pressure.

Table 3: ROP versus pressure drop for UBD wells

Reservoir pressure (psi)	Pressure drop (ΔP) (psi)	Rate of penetration (ft/h)	Lithology
2900	290	45	Sandstone
3000	360	38	Sandstone
1350	540	16	Sandstone
3200	640	27	Sandstone
5500	990	30	Sandstone

Table 4 gives a recorded data for different wells drilled by aerated fluid in a reservoir that has a constant pressure and same lithology compared to those wells drilled in overbalanced environment (Moore and Lafave, 1956).

Table 4: Recorded data for UBD wells

Pressure drop (psi)	ROP (ft/h)	Production while drilling (%)	Production after test (%)	Lithology
290	26.6	0	1.2	Sandstone
320	44.7	0.8	3.9	Sandstone
350	19.45	1	2	Sandstone
406	22.5	1.5	1.8	Sandstone
435	17.6	2.7	3.4	Sandstone

Fig. 7 illustrates that ROP initially decreases with an increase in pressure drop and increases with further increase in pressure drop. Whereas, Fig. 8 shows that ROP has no definite relation with pressure drop if other drilling parameters are ignored. However a continuous increase in formation fluid production while drilling was observed with the continuous increase in pressure drop as shown in Fig. 9.

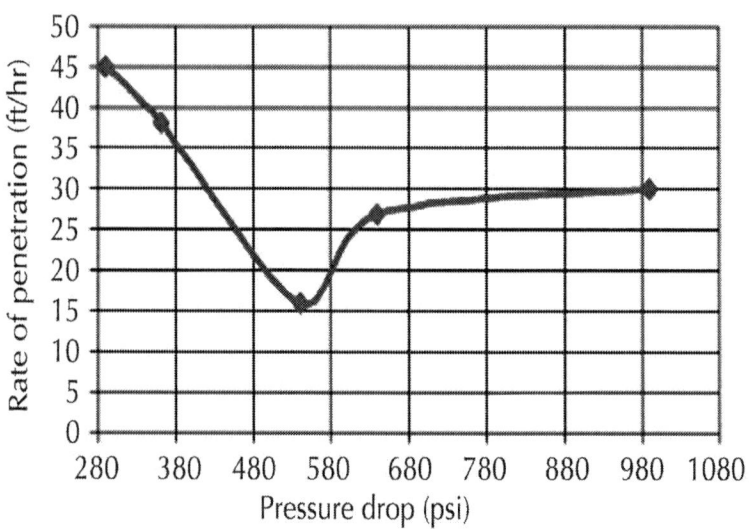

Figure 7: ROP versus pressure drop for UBD wells in different reservoirs.

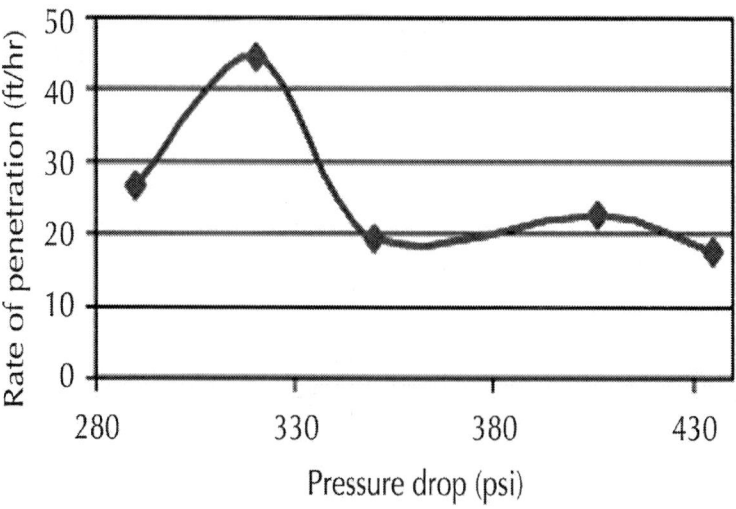

Figure 8: ROP versus pressure drop for UBD wells in one reservoir.

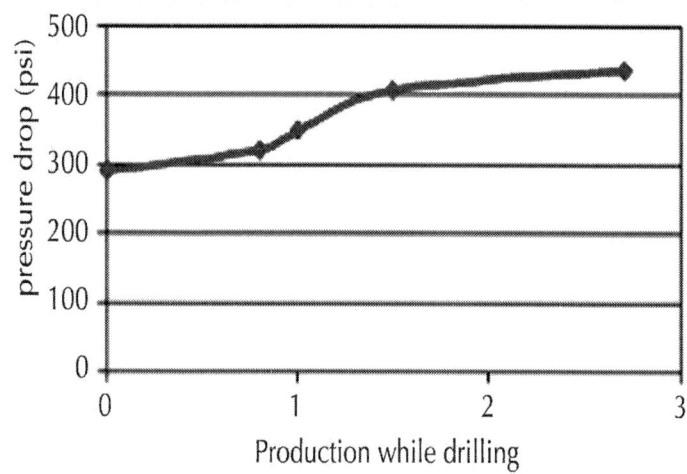

Figure 9: Production while drilling versus pressure drop for UBD wells.

Fig. 10 illustrated that all wells drilled by UBD have an increased in fluid production rate compared to those wells drilled in overbalanced environment. In addition, there is no clear relation between the amount of fluid production while drilling and the amount of fluid production after the well is put on production as shown in Fig. 10.

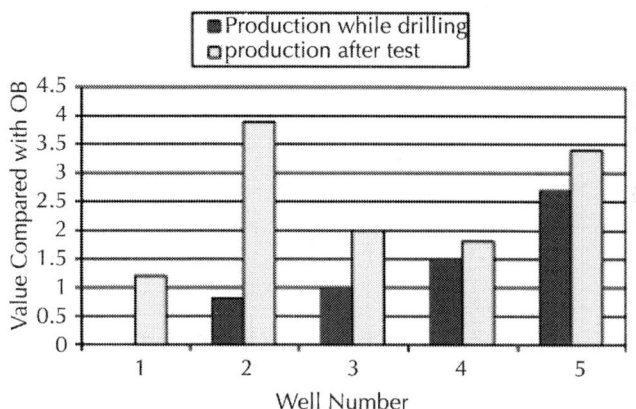

Figure 10: Comparison of production while and after UBD drilling.

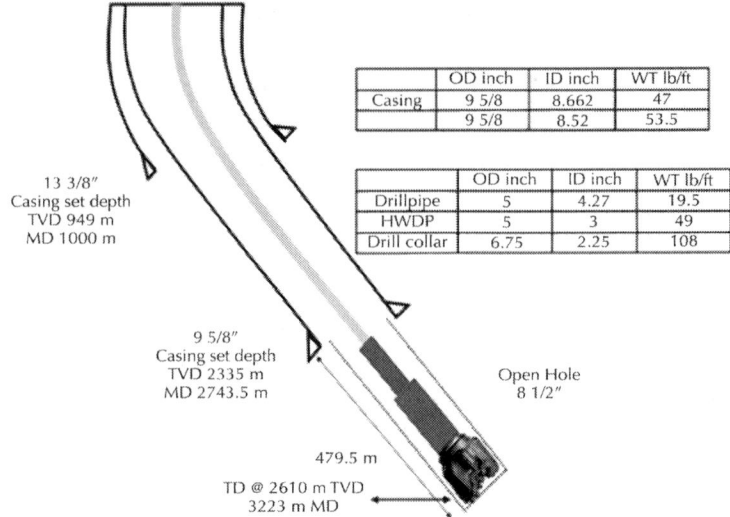

		OD inch	ID inch	WT lb/ft
Casing		9 5/8	8.662	47
		9 5/8	8.52	53.5

	OD inch	ID inch	WT lb/ft
Drillpipe	5	4.27	19.5
HWDP	5	3	49
Drill collar	6.75	2.25	108

13 3/8"
Casing set depth
TVD 949 m
MD 1000 m

9 5/8"
Casing set depth
TVD 2335 m
MD 2743.5 m

Open Hole
8 1/2"

479.5 m

TD @ 2610 m TVD
3223 m MD

Figure 11: Well schematic of Gulf of Suez oil field area.

Table 5 highlights the savings in total rig days and cost for conventional versus underbalanced drilling wells in Iran (Roving and Reynolds, 1994). It is clear that big savings in drilling cost was realized.

Table 5: Drilling time and cost savings for 8-1/2" hole section drilled underbalanced conditions

Well	Real cost		Clean cost (just drilling)	
	Days	K$	Days	K$
8-1/2" hole – conventional				
1	27	1171	27	1171
2	25.7	1146.3	24.4	1114
3	30.4	2125.3	21.6	1771.9
4	19.3	1360.1	17.6	1230.8
5	31.9	2215.7	16.7	1629.3
6	23.3	1058.5	22.4	1035
7	31.4	1385.1	23	1005.6

8	21.6	1241.5	17.8	989.9
9	20.7	899.1	17.2	667.4
10	34.1	1551.6	30.3	1300.1
Average	26.5	1415.4	21.8	1191.5

The cost savings ranged between $90,000 and $110,000 for 8-1/2 in. hole section and between $170,000 and $190,000 for the 6-1/2 in. hole size (Table 6). A total of approximately $1.4MM has been saved (drilling only) and about $1MM (overall), for the five wells drilled.

Table 6: Drilling time and cost savings for 6-1/2″ hole section drilled underbalanced conditions

Well	Total cost		Drilling cost	
	Days	K$	Days	K$
6-1/2″ hole – conventional				
1	9	886.6	9	886.6
2	11.8	591.8	11.8	591.8
3	20.7	1186.4	18.1	1082
4	29.6	1596.7	17.8	644.7
5	33.5	2074.1	20	1531.9
6	21.9	928.1	19.7	779.9
7	19.1	995.5	17.8	938.3
8	14.1	778.5	11.8	650.6
9	16.4	800.8	16.4	800.8
Average	19.6	1093.2	15.8	878.5
6-1/2″ hole – underbalanced				
1	7.4	507.8	6.6	471.9
2	24	1664.6	11.9	998.9
3	22.4	1804	17.2	1057.7
4	14.8	545.1	10.8	387.57
5	9.5	580.6	9	560.6
Average	15.6	920.4	11.1	695.3

PROPOSED UBD PROGRAM TO BE IMPLEMENTED IN EGYPTIAN FIELDS

Based on the experience and the problem faced discussed in the previous discussions, a proposed UBD program is given here-below.

Gulf of Suez Oil Field Area

The selected example includes drilling through the reservoir section, which consists of two production formations (Belayim and kareem formation from Miocene age). The reservoir and formation characteristics are given in Table 7 and Table 8.

Table 7: Gulf of Suez reservoir characteristics

Parameter	Belayim	Kareem
Pressure	1500 psi	1700 psi
Temperature	180 °F	190 °F
Gas–oil ratio (GOR)	15–17 SCF/STB	20 SCF/STB
Porosity (md)	18–20%	20–22%
Permeability	200 md	500 md
API⁰ gravity of oil	20–23	20–30
H_2S concentration	No	No

Table 8: Gulf of Suez formation characteristics

Formation	Lithology	Top (m)	Thickness (m)	Pore pressure (psi)
Belayim				
Hammam Faraun	Shale-sand	2160	35	

Ferran	Shale-sand	2195	140	
Sidri	Mainly sand	2335	65	1500
Babaa	Anhydrite	2400	15	
Kareem	Limestone	2415	195	1700

The selected reservoir can be drilled by underbalanced drilling technique and the proposed UBD program is given in Table 9. Fig. 12 shows the operating window, multiphase fluid injection of Gulf of Suez oil field area.

Table 9: Underbalanced drilling design parameters for Gulf of Suez area

Rig modification	• No essential modifications to be made on the rig to suite UBD operations • The substructure has to be high enough to allow Rotating Control Head (RCH) to be installed on top of the Hydril
Well plan	• As shown in Fig. 11
Drill string design	• Use a 5" DP and 5" HWDP on 6-3/4" DC
BHA	• The BHA consists of 6-1/2" mud motor and MWD to drill 8-1/2" hole • An 8-1/2" bit size of 3 × 13/32" nozzles
Drilling fluid selection	• The deviated section will be drilled using an oil bas mud and a membrane nitrogen generation circulating system
A-liquid phase	• Drilling fluid is native crude oil with density 7.6 ppg (0.91 S.G. or 20° API) • Liquid flow rates were selected to achieve a drawdown from the reservoir pressure
B-gas phase	• Nitrogen was selected as the injection gas • •Nitrogen will be obtained from the surrounding air and generated onsite

Operating envelope	• A minimum drawdown at the bit of 100 psi is required to ensure adequate under-balanced conditions in the well • Using 300 gpm and more than 2400 scfm of Nitrogen will provide maximum 100 psi drawdown from the expected reservoir pressure, as shown in Fig. 12 • In case the real reservoir pressure will result below the expected value, then the liquid injection rate should be reduced increasing the risk for a hole cleaning issue
Hole cleaning	• Minimum annular liquid velocities in deviated holes of 210 ft/min when crude oil is used as the drilling fluid to ensure that the drilled cuttings are effectively removed from the wellbore • A wiper drilling trip will help clear the problem of hole cleaning
Motor performance	• The motor should be suitable for oil/nitrogen two-phase application • A maximum Equivalent Liquid Volume through the motor of 600 gpm was used as reference • A pressure loss of 800 psi between downhole motor and MWD was considered • The motor should not have a bypass valve on top of it
Production sensitivity	• As more reservoir fluids (oil and gas) introduced into the wellbore, the bottomhole circulating pressures (BHCP) will decrease • BHCP will therefore be controlled by increasing liquid injection and/or decreasing nitrogen injection, based on real-time BHCP data from the MWD tool • BHCP could also be controlled with surface backpressure • Choking will be necessary in stabilizing the circulating system during and after drill string connections

Data acquisition	• The software for the rig data acquisition has to be able to interface with the UBD equipment software
Completion	• The well can be completed with barefoot completion technique, or installing a slotted liner completions

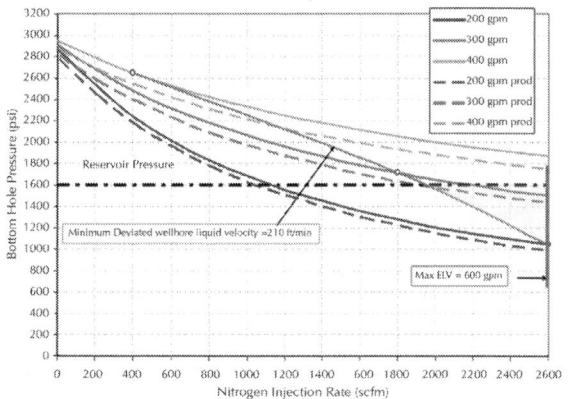

Figure 12: Operating window, multiphase fluid injection of Gulf of Suez oil field area.

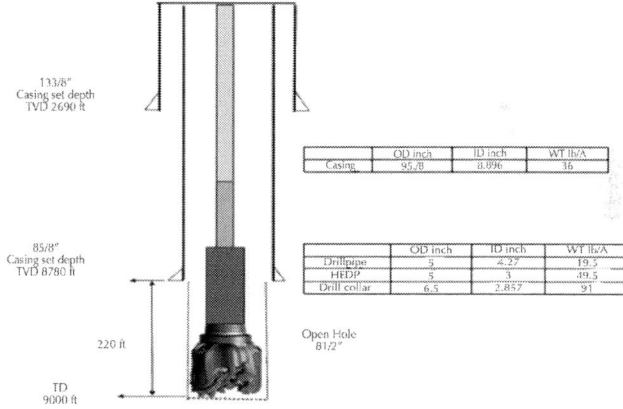

Figure 13: Well schematic of western desert oil field area.

Western Desert Oil Field Area

The selected example includes drilling through the reservoir section, which consists of Alam El Buieb formation of Cretaceous age. The lithology of this formation is sandstone with depleted reservoir pressure 1600 psi, reservoir temperature 219 °F, porosity 19%, permeability 200 md, GOR 95 SCF/STB, 41.7° API gravity of oil, and there is no H_2S concentration. The selected reservoir can be drilled by underbalanced drilling technique as given in Table 10. Fig. 14 shows the operating window, multiphase fluid injection of western desert oil field area.

Table 10. Underbalanced drilling design criteria for western desert area

Rig modification	• No essential modifications to be made on the rig to suite UBD operations • The substructure has to be high enough to allow Rotating Control Head (RCH) to be installed on top of the Hydril
Well plan	• As shown in Fig. 13
Drill string design	• Use 5" DP, 5" HWDP and 6.5" DC
BHA	• No downhole motor used • An 8-1/2" bit size of 3 × 13/32" nozzles size
Drilling fluid selection	• Based on the pore pressure and formation depth, the reservoir formation is below the normal pressure regime • The subnormal pressure requires the use of a multi-phase (liquid + gas) drilling fluid system in order to obtain on Underbalanced drilling condition
A-liquid phase	• Drilling fluid is native crude oil with density 6.84 ppg (0.82 S.G. or 41.7° API) • Liquid flow rates were selected to achieve a draw-down from the reservoir pressure
B-gas phase	• Nitrogen was selected as the injection gas
Operating envelope	• It is displayed as the area of the graph between the targets BHCP's, bound by the maximum motor through-put, the minimum annular liquid velocity, Fig. 11 • Using 300 gpm and more than 2200 scfm of Nitrogen will provide maximum 200 psi drawdown from the expected reservoir pressure

Hole cleaning	<ul><li>Depends on several variables such as cutting size and shape; liquid properties; drill string rotation; liquid velocities; flow regime, etc.</li><li>Minimum vertical annular liquid velocities of 180 ft/min when crude oil is used as the drilling fluid to ensure that the drilled cuttings are effectively removed from the wellbore</li></ul>
Hydraulic modeling	<ul><li>Using a multiphase hydraulic simulator, the required underbalanced drilling parameters could be evaluated in detail</li><li>Graphs can be created to incorporate the limiting factors of minimum annular liquid velocity required for hole cleaning and the desired BHCP range</li></ul>
Pressure while drilling	<ul><li>When the maximum gas volume fraction (GVF) inside the drill pipe is bellow, 20% conventional mud pulse tools (MWD/LWD/PWD) can be used</li><li>Otherwise, electromagnetic transition tools have to be used in order to obtain downhole data real time</li></ul>
Data acquisition	<ul><li>The software for the rig data acquisition has to be able to interface with the UBD equipment software</li></ul>
Completion	<ul><li>The well can be completed with barefoot completion technique, or installing a slotted lined</li></ul>

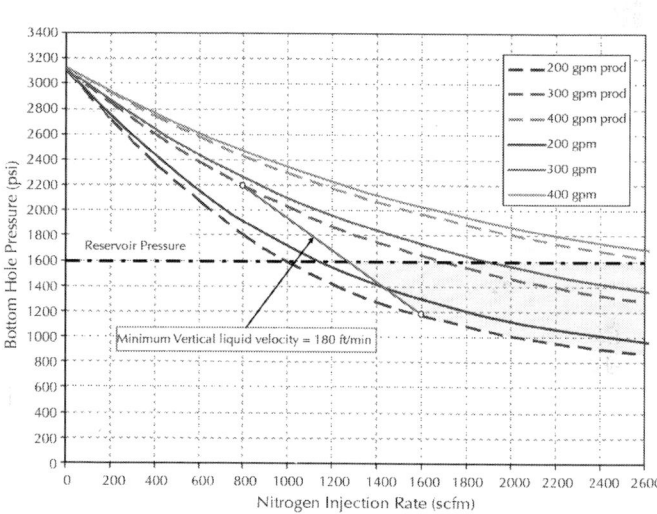

Figure 14: Operating window, multiphase fluid injection of western desert oil field area.

Nile Delta Oil Field Area

The selected example includes the reservoir section, which consists of one production formation (Qawasim from Miocene age). It has a sandstone lithology with reservoir pressure 3800 psi, reservoir temperature 185 °F, GOR 1100 SCF/STB, average porosity 25%, average permeability 400 md, gravity of oil 50° API, and there is no H_2S concentration.

The selected reservoir can be drilled by underbalanced drilling technique as given in Table 11. Fig. 16 shows the operating window, multiphase fluid injection of nile delta oil field area.

Table 11: Proposed UBD program in Nile Delta area

Rig modification	<ul><li>No essential modifications to be made on the rig to suite UBD operations</li><li>•The substructure has to be high enough to allow Rotating Control Head (RCH) to be installed on top of the Hydril</li></ul>
Well plan	As shown in Fig. 15
Drill string design	<ul><li>Use a 5" DP, 5" HWDP and 6.5" DC</li><li>An 8-1/2" bit size of 3x13/32" nozzles</li></ul>
BHA	<ul><li>The BHA consists of 6-1/2" PDM mud motor and MWD to drill 6" hole</li><li>If MWD signal doesn't observed, use electromagnetic MWD tools</li></ul>
Drilling fluid selection	<ul><li>Water based fluid (flow-drilling operation)</li><li>Drilling fluid is water with density 8.75 ppg (1.05 S.G.)</li><li>Liquid flow rates and surface choke backpressure were selected to achieve a drawdown from the reservoir pressure</li></ul>

Operating envelope	• It is recommended to pump at least 400 gpm of liquid phase to avoid any operational problem related with hole cleaning • The drawdown is 200 psi to prevent wellbore collapse
Motor perfor-mance	• A maximum equivalent liquid volume through the motor of 600 gpm was used as reference • A pressure loss of 800 psi between downhole motor and MWD was con-sidered
Hole cleaning	• Minimum annular liquid velocities in deviated holes of 180 ft/min to ensure that the drilled cuttings are effectively removed from the wellbore • A wiper trip will help clear the hole cleaning problem
Tripping	• Some type of snubbing device can be used, or a downhole isolation valve can be installed • Balancing the well for trips seemed the simplest and least expensive method
Data acquisi-tion	• The software for the rig data acquisi-tion has to be able to interface with the UBD equipment software
Completion	• The well can be completed with bare-foot completion technique, or install-ing a slotted lined

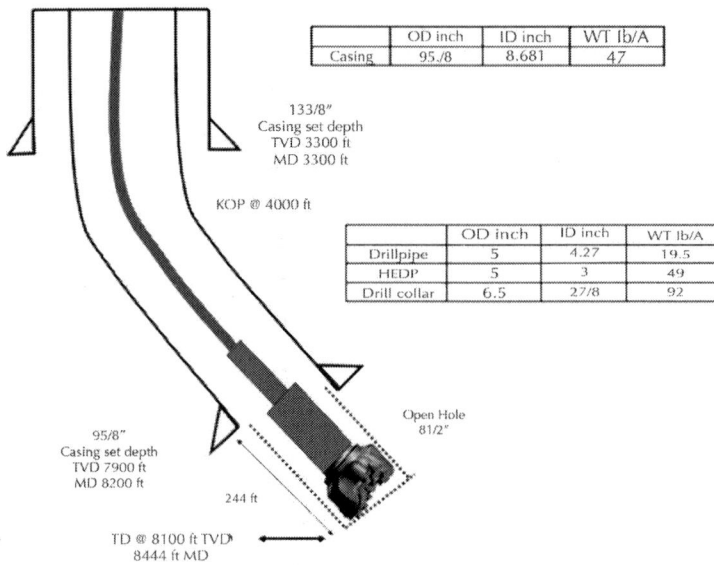

Figure 15: Well schematic of Nile delta oil field area.

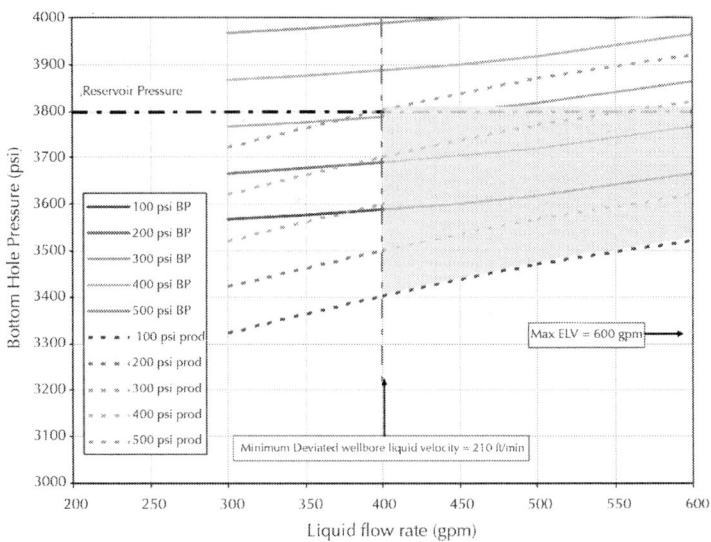

Figure 16: Operating window, flow-drilling operation for Nile delta oil field area.

CONCLUSIONS

Planned and applied correctly, underbalanced drilling technology can address problems of formation damage, lost circulation and poor penetration rates. The ability to investigate and characterize the reservoir while drilling is another important benefit of under balanced drilling. Based on the analysis of the real cases studied during the research, the following conclusions could be cited:

- Underbalanced drilling technique is a very useful technique especially when applied in reservoir section. It prevents formation damage, increases ROP, increases reservoir productivity and reduces the total cost of the well.

- Candidate screening is a rigorous and is a critical first step in the design of a successful underbalanced drilling operation. Although UBD has many advantages, it is not a magic solution for all fields or drilling problems. Poor screening and planning would result in an over-enthusiastic misapplication of the technology, and possibly failure.

- Many issues must be considered when designing an underbalanced drilling project including but certainly not limited to rock properties, reservoir pressure, borehole stability, drilling fluid type, injection method for gas assist, effect of compressible fluid on MWD, downhole motor requirements, bit type, corrosion, equipments availability, separation and fluid handling requirements especially when dealing with hydrocarbon drilling fluid, tripping procedures, data acquisition and completion procedures. Proper planning and design work, addressing these parameters, is essential to successfully conduct an underbalanced drilling project.

- UBD with stable foam through depleted reservoirs can be conducted safely and successfully in both vertical and horizontal wells. Drilling with foam has some appeal because foam has some attractive qualities and properties with respect to the very low hydrostatic densities, which can be generated with foam systems. Foam has good rheology and excellent cutting transport properties.

- Real time capture of production data while drilling should provide information about the reservoir not otherwise available.
- A proposed UBD program to be implemented in Egyptian fields is developed.

REFERENCES

1.   Azeemddin, M. et al., 2006. Underbalanced Drilling Borehole Stability

2.   and Implementation in Depleted Reservoirs, a Joaquin Field, Eastern Venezuela. IADC/SPE99165, February, 2006.

3.   Bates, R.E., 1965. Field Results of Percussion Air Drilling. SPE 886, March, 1965.

4.   Bennion, D.B., Thomasand, F.B., Bietz, R.F., 1998. Underbalanced Drilling: Praises and Perils. SPE Drilling and Completion, December, 1998.

5.   Bentsen, N.W., Veny, J.N., 1976. Preformed Stable Foam Performance in Drilling and Evaluating Shallow Gas Wells in Alberta. SPE 5712-PA, Formation Damage Conference held in Houston, October, 1976.

6.   Black, A.D., Green, S.J., 1978. Laboratory simulation of deep well drilling. Pet. Eng., 40.

7.   Bourgoyne, A.T., Young Jr., F.S., 1974a. A multiple regression approach to optimal drilling and abnormal pressure detection. SPEJ, 371.

8.   Bourgoyne, A.T., Young Jr., F.S., 1974b. A multiple regression approach to optimal drilling and abnormal pressure detection. Trans. AIME, 257.

9.   Boyun, Guo, Rajtar, J.M. 1995. Volume Requirements for Aerated Mud Drilling. SPE 26956-PA, Drilling and Completion, California Regional Meeting held in Ventura, September, 1995.

10.   Claytor, S.B., Manning, K.J., Schmalzried, D.L., 1991. Drilling a Medium-radius Horizontal Well with Aerated Drilling Fluid: A

11.   Case Study. Paper SPE 21988 presented at the 1991 SPE/IADC Drilling Conference, Amsterdam, March 11–14.

12.   Cunningham, R.A., Eenink, J.G., 1959. Laboratory study of effect of overburden, formation, and mud column pressures on drilling rate

13.   permeable formations. Trans. AIME 216, 9.

14.   Dorenbos, Roelien, Ranalho, Jone, 2002. Underbalanced Drilling Primer. Shell International Exploration and Production B.V., June, 2002.

15.   Eckel, J.R., 1957. Effect of pressure on rock drillability. Trans. AIME 213, 1. Gamier, A.J., van Lingen, N.H., 1959. Phenomena affecting drilling rates at depth. Trans. AIME 216, 232.

16.   George, E., Waston, Ralpha, A., 1956. Review of Air and Gas

17.   SPE 703-G, Petroleum Branch Fall Meeting in Los mAngeles, October, 1956.

18.   Godwin, A., Lokpobiri, Ikoku, Chi U., 1986. Volumetric Requirements for Foam and Mist Drilling Operations. SPE 11723-PA, Petroleum Branch Office, California Regional Meeting held in Ventura, February, 1986.

19.   Gordon, D. et al., 2005. Underbalanced Drilling with Casing Evolution in the south Texas Vicksburg. SPE Drilling and Completion, June, 2005.

20.   Gray, Kenneth E., 1957. The Cutting Carrying Capacity of Air at Pressure above Atmospheric. SPE 874-G, October, 1957. Hongren, G.U., Walton, J.C., Stein, D.A., 1999. Designing under- and near-balanced coiled-tubing drilling by use of computer simulations. SPE Dril. Comp. 14 (2).

21.   Hooshmandkoochi, A., Zaferanich, M., Malekzadeh, A., 2007. First Application of Underbalanced Drilling in Fractured Carbonate Formations of Iranian Oil Fields Leads

to Operational Success and Cost Saving. SPE 105536-MS, Middle East Oil and Gas Conference

22. held in Bahrain International Exhibition Center, Kingdom of Bahrailn, March, 2007.

23. International Association of Drilling Contractor, 2005. IADC Well Classification System for Underbalanced Operations and Managed Pressure Drilling <http://www.iadc.org/committees/underbalanced/>, March, 2005.

24. Kuru, E. et al., 1999. New Directions in Foam and Aerated Mud Research and Development. SPE 53963-MS, Latin American

25. Petroleum Engineering Conference held in Caracas, Venezuela, April, 1999.

26. Louison, R.F., Reese, R.T., Andrews, J.P., 1984. Case History: Underbalance Drilling the Midway and Navarro Formations Successfully in Hallettsville, TX. SPE13112, September, 1984.

27. Maclovio, Yanez M., 1996. PEP Region Norte and Valenzuela J.

28. Marten, Tecominoacan 408: First Underbalance application in MEXECO. SPE 35320, March, 1996.

29. Maurer Engineering Manual, 1998. Underbalanced Drilling and Completion Manual, November, 1998.

30. Meng, Y. et al., 2005. Discussion of Foam Corrosion Inhibition in Air Foam Drilling. SPE 94469-MS, International Symbosium on Oil Field Corrosion held in Aberdeen, United Kingdom, May, 2005.

31. Moore, C.L., Lafave, V.A., 1956. Air and Gas Drilling. SPE 494-G, February 1956.

32. Moore, D.D., Bencheikh, A., Chopty, J.R., 2004. Drilling Underbalanced in Hassi Messaud. SPE/IADC 91519, October, 2004.

33. Murray, A.S., Cunningham, R.A., 1955. Effect of mud column pressure on drilling rates. Trans. AIME 204, 196.

34. Nas, S., 2004. Leading Edge Advantage Ltd – Introduction to Underbalanced Drilling Manual, February, 2004.

35. Negra, A.F., Lage, A.C.V.M., Cunha, J.C., 1999. An Overview of Air/Gas/Foam Drilling in Brazil. SPE 56865-PA, Drilling and Completion 14 (2), Drilling Conference held in Amsterdam, June, 1999.

36. Parra, J.G., Cells, E., Gennare, S., 2003. Wellbore Stability Simulations or Underbalanced Drilling Operations in Highly Depleted Reservoirs. SPE Drilling and Completion, June, 2003.

37. Qutob, H.H. et al., 2007. The Successful Application of Underbalanced Drilling Technology for Reservoir Evaluation and Drilling Performance Improvement in Kuwait. SPE 106680, June, 2007.

38. Qutob, Hani, Ferreira, Horacio, 2005. The SURE way to Underbalanced Drilling. SPE 93346, March, 2005.

39. Rankin, M.D., Friesenhahn, T.J., Price, W.R., 1989. Lightened Fluids Hydraulics and Inclined Bore Holes. Paper SPE 18670 presented at the 1989 SPE/IADC Drilling Conference, New Orleans, Feb. 28– March 3.

40. Roving, J.W., Reynolds, E., 1994. Underbalanced Drilling Through Oil Production Zones With Stable Foam in Oman. IADC/SPE 27525, February, 1994.

41. Salah El-Din, M.A., El-Katatney, S.M. (2009). Implementation of Underbalanced Drilling Technique in Egyptian Fields. M.Sc.

42. Thesis, Cairo University, Egypt, 2009. Sunthankar, A.A. et al., 2001. New Developments in Aerated Mud Hydraulics for Drilling in Inclined Wells. SPE67189, March, 2001.

43. Vidrine, D.J., Benit, E.J., 1968. Field verification of the effect of differential pressure on drilling rate. JPT, 676.

44. Weatherford Company, 2006. Operational Sequence in UBD (ROAD MAP). Weatherford Controlled Pressure Drilling and Testing Services.

45. Westermark, R.V., 1986. Drilling with a Parasite Aerating String in the Disturbed Belt, Gallatin County, Montana. IADC/SPE 14734, February, 1986.

46. Whiteley, Maxwel C., England, William P., 1986. Air Drilling Operation Improved by Percussion-Bit/Hammer-Tool Tandem. SPE Drilling Engineering, October, 1986.

47. Zhou, L. et al., 2005. Hydraulics of Drilling with Aerated Mud under Simulated Borehole Conditions. SPE/IADC 92484, February, 2005.

Chapter 7

Fractures and Fracturing: Hydraulic Fracturing in Jointed Rock

Charles Fairhurst[1, 2]

[1]Senior Consultant, Itasca Consulting Group, Inc, Minneapolis Minnesota, USA

[2]Professor Emeritus, University of Minnesota, Minneapolis, Minnesota, USA

ABSTRACT

Rock in situ is arguably the most complex material encountered in any engineering discipline. Deformed and fractured over many millions of years and different tectonic stress regimes, it contains fractures on a wide variety of length scales from microscopic to tectonic plate boundaries.

Hydraulic fractures, sometimes on the scale of hundreds of meters, may encounter such discontinuities on several scales. Developed initially as a technology to enhance recovery from petroleum reservoirs, hydraulic fracturing is now applied in a variety of subsurface engineering applications. Often carried out at depths of kilometers, the fracturing process cannot be observed directly.

Early analyses of the hydraulic fracturing process assumed that a single fracture developed symmetrically from the packed off-pressurized interval of a borehole in a stressed elastic continuum. It is now recognized that this is often not the case. Pre-existing fractures can and do have a significant influence on fracture development, and on the associated distributions of increased fluid pressure and stresses in the rock.

Given the usual lack of information and/or uncertainties concerning important variables such as the disposition and mechanical properties of pre-existing fracture systems and properties, rock mass permeabilities, in-situ stress state at the depths of interest, fundamental questions as to how a propagating fracture is affected by encounters with pre-existing faults, etc., it is clear that design of hydraulic fracturing treatments is not an exact science.

Fractures in fabricated materials tend to occur on a length of scale that is small; of the order of the 'grain size' of the material. Increase in the size of the structure does not introduce new fracture sets.

Numerical modeling of fracture systems has made significant advances and is being applied to attempt to assess the extent of these uncertainties and how they may affect the outcome of practical fracturing programs. Geophysical observations including both micro-seismic activity and P- and S-wave velocity changes during and after stimulation are valuable tools to assist in verifying model predictions and development of a better overall understanding of the process of hydraulic fracturing on the field scale. Fundamental studies supported by laboratory investigations can also contribute significantly to improved understanding.

Given the widening application of hydraulic fracturing to situations where there is little prior experience (e.g., Enhanced Geothermal Systems (EGS), gas extraction from 'tight shales' by fracturing in essentially horizontal wellbores, etc.) development of a greater understanding of the mechanics of hydraulic fracturing in naturally fractured rock masses should be an industry-wide imperative. HF 2013 International Conference for Effective and Sustainable Hydraulic Fracturing is very timely!

This lecture will describe examples of some current attempts to address these uncertainties and gaps in understanding. And, it is hoped, it will stimulate discussion of how to achieve more effective practical design of hydraulic fracturing treatments.

INTRODUCTION

The term 'rock' covers a wide variety of materials and widely different rheological properties often proximate to each other in the subsurface. Tectonic and gravitational forces, sustained over millions of years, have deformed and fractured the rock on many scales. These forces are transmitted in part through the solid skeleton of the rock, and in part through the fluids under pressure in the pore spaces. Long-term circulation through rock at high temperatures at depth involves dissolution and precipitation along the fluid pathways, producing changes in the chemical composition of the fluids and modifying the overall fluid circulation.

Rock in situ is 'pre-loaded' and in a state of changing equilibrium. Any engineering activity changes this equilibrium. Often the changes can be accommodated in stable fashion, but serious instabilities can develop.

The rock mass is opaque. Although geophysics is making impressive advances in defining large structures such as faults and bedding planes, most of the features that influence the rock response to engineering activities remain hidden. Mining and civil engineering activities allow three-dimensional access to the underground and direct observation of smaller features such as fracture networks, but most of the newer engineering applications involve essentially one-dimensional access by borehole. Rock engineering problems fall into the 'data –limited' category, as defined by Starfield and Cundall (1988), and strategies to address them must follow a different strategy than engineering problems where detailed and precise design information is available.

Faced with such complexity and lack of structural details, traditional subsurface engineering design has been guided by empirical procedures developed and refined through long experience.

Projects are now venturing well beyond current experience, and for many, 'novel' applications now considered (e.g., Enhanced Geothermal Systems, Carbon Sequestration, see Appendix 1). There is little experience, few guiding rules and very little data to guide the engineering approach.

Such obstacles notwithstanding, subsurface processes, both long–term geological and short term responses, to engineering activities do obey the laws of Newtonian Mechanics.

Classical continuum mechanics has long been used to guide some aspects of design, but considerable care is required in practical application, due to the need to simplify the representation of the real conditions in order to obtain analytical solutions.

The remarkable developments in high-speed computation and associated modeling techniques over the past one to two decades provide an important new tool, which complemented by the appropriate field instrumentation, can augment the classical continuum analyses and help overcome the lack of prior experience. Some empiricism and general practical guidelines may still be useful for the design engineer, but these can and should be mechanics-informed.

This lecture attempts to illustrate the 'mechanics-informed' approach with respect to the practical application of hydraulic fracturing and related engineering procedures to rock engineering.

HYDRAULIC FRACTURING

Hydraulic fracturing first was used successfully in the late 1940's to increase production from petroleum reservoirs (Howard and Fast, 1970). The technology has evolved since and is now a major, essential technique in oil and gas production. This and other impressive oil industry developments, such as directional drilling, have attracted interest in application of these technologies to a variety of other subsurface engineering operations. Enhanced Geothermal Energy (EGS) is a notable example. Geothermal Energy is a huge resource. Commenting on the EGS resource in the USA, Tester et al. (2005), state:

"....we have estimated the total EGS resource base to be more than 13 million exajoules (EJ)[1] - . Using reasonable assumptions regarding how heat would be mined from stimulated EGS reservoirs, we also estimated the extractable portion to exceed 200,000 EJ or about 2,000 times the annual consumption of primary energy in the United States in 2005. With technology improvements, the economically extractable amount of useful energy could increase by a factor of 10 or more, thus making EGS sustainable for centuries." [2] -

"At this point, the main constraint is creating sufficient connectivity within the injection and production well system in the stimulated region of the EGS reservoir to allow for high per-well production rates without reducing reservoir life by rapid cooling." [3] -

Field experiments to extract geothermal energy from rock at depth by hydraulic fracturing were started in 1970 by scientists of the Los Alamos National Laboratory, USA. Two boreholes were drilled into crystalline rock (one 2.8 km deep, rock temperature 195°C; the other 3.5 km rock, 235°C) at Fenton Hill, New Mexico. Hydraulic fracturing was used to develop fractures from the boreholes in order to create a fractured region through which water could be circulated to extract heat from the rock. The experiment was terminated in 1992. Commenting on what was learned from the Fenton Hill study, Duchane and Brown (2002) note: "The idea that hydraulic pressure causes competent rock to rupture and create a disc-shaped fracture was refuted by the seismic evidence. Instead, it came to be understood that hydraulic stimulation leads to the opening of existing natural joints that have been sealed by secondary mineralization. Over the years additional evidence has been generated to show that the joints oriented roughly orthogonal to the direction of the least principal stress open first, but that as the hydraulic pressure is increased, additional joints open."

This is an early indication that pre-existing fractures mass significantly affect how hydraulic fractures propagate in a rock mass.

INFLUENCE OF FRACTURES AND DISCONTINUITIES ON THE STRENGTH OF BRITTLE MATERIALS

Hydraulic fracturing can be considered as a technique to overcome the strength of a rock mass in situ, initiation and propagation of a crack through a system of pre-existing fractures, essentially planar discontinuities (e.g., bedding planes), and intact rock.

In examining the fracture propagation process, the pioneering work of Griffith (1921, 1924) is a logical point of departure. Griffith had identified planar discontinuities, or flaws, in fabricated materials as the reason why the observed technical strength of brittle materials was about three orders of magnitude lower than the theoretical inter-atomic cohesive (tensile) strength.[4] - Using an analytical solution byInglis (1913) for the elastic stresses generated around an elliptical crack in a plate, Griffith observed that the maximum tensile stress at the tip of the crack $_t = _0 (1 + 2a/b)$, where a and b are the major and minor semi-axes of the ellipse, and as the ellipse degenerated to a sharp crack or flaw (i.e., as the ratio a/b became very high)[5] - , the stress $_t$ could rise to a value high enough to reach the inter-atomic cohesive strength sufficient to cause the original crack to start to extend.

But would the crack continue to extend and lead to macroscopic failure? To address this question, Griffith invoked the *Theorem of Minimum Potential Energy,* which may be stated as "The stable equilibrium state of a system is that for which the potential energy of the system is a minimum." For the particular application of this theorem to brittle rupture, Griffith added the statement, "The equilibrium position, if equilibrium is possible, must be one in which rupture of the solid has occurred, if the system can pass from the unbroken to the broken condition by a process involving a continuous decrease of potential energy."[6] -.

Griffith's classical work has provided the foundation for the field of "Fracture Mechanics" [Knott (1973); Anderson (2005)] responsible for major continuing advances in the development of high-performance fabricated materials.

Since we will make reference later to this specific definition by Griffith, it is useful to re-state it here.

THEOREM OF MINIMUM POTENTIAL ENERGY

"The stable equilibrium state of a system is that for which the potential energy of the system is a minimum. The equilibrium position, if equilibrium is possible, must be one in which rupture of the solid has occurred, if the system can pass from the unbroken to the broken condition by a process involving a continuous decrease of potential energy."

Although much of classical Fracture Mechanics has emphasized applications to problems of Linearly Elastic Fracture Mechanics (LEFM) it is important to recognize that the theorem of minimum potential applies equally to inelastic problems.

MECHANICS OF HYDRAULIC FRACTURING

As used classically in petroleum engineering, hydraulic fracturing involves sealing off an interval of a borehole at depth in an oil or gas bearing horizon, subjecting the interval to increasing fluid pressure until a fracture is generated, injecting some form of granular proppant into the fracture as it extends a considerable distance from the borehole into the petroleum bearing formation, and then releasing the pressure. This causes the sides of the fracture to compress onto the proppant, creating a high-permeability pathway to allow oil and/or natural gas to flow back to the well and to the surface.

Figure 1 shows a simple two-dimensional cross-section through an idealized hydraulic fracture. The borehole injection point is at the center of the fracture, which is assumed to be a narrow ellipse that has extended in a plane normal to the direction of the maximum[7] - (least compressive) in-situ stress.

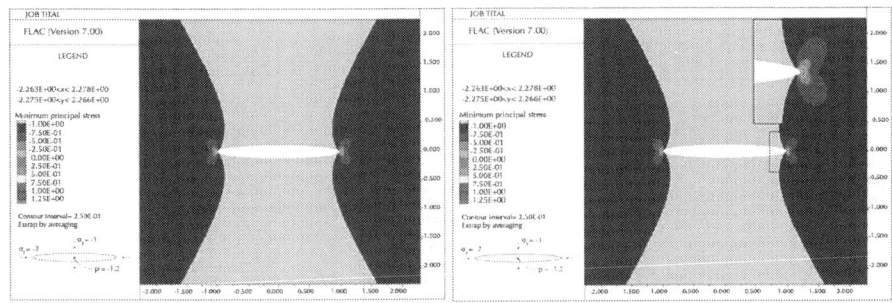

Figure 1: Left) Major and (right) minor principal stresses in the vicinity of an internally pressurized elliptical crack in an impermeable rock.

In the case shown, the crack major/minor axis ratio a/b is 10:1. The internal fluid pressure $p = 1.2$, while the least compressive principal stress $x = 1.0$. This results in a tensile stress concentration at the crack tip. The magnitude of the elastic stress concentration at the crack tip increases directly with 2a/b, (Inglis, 1913). Hence for the case of a>>b, i.e., a 'sharp' crack [8] - , the concentration is very high, and the crack will extend essentially as soon as the fluid pressure exceeds the magnitude of the least compressive principal stress (x in Figure 3) it begins to extend, and there will be a pressure gradient from the injection point towards the crack tip as the fluid flows towards the tips. This gradient will depend on the fluid viscosity. Also, since the rock will exhibit some level of permeability, fluid will also flow (or 'leak–off') into the formation as it flows under pressure along the fracture; the rock has a finite strength, or 'toughness' so that energy will be required to extend the crack.

An analytical solution for the stresses in the elastic medium and the crack-opening displacement along the crack was first published by Inglis (1913) and served as the basis for early applications to hydraulic fracturing and fracture treatment design. The Perkins, Kern (1961) and Nordgren (1972) (PKN) andGeertsma and de Klerk (1969) (GDK) models are still used, although numerical models and combinations are now popular. Details of the PKN and GDK models can be found on the SPE website: http://petrowiki.spe.org/Fracture_propagation_models. Several differences between the stationary crack assumed by Inglis (1913) and a hydraulic fracture introduce significant difficulties in developing an accurate model of the fracturing process. Thus, the fracture is generated by application of an increasing fluid pressure until the fracture is initiated and extends away from the injection point. Flow of fluid in the fracture is governed by classical fluid flow equations of Poiseuille and Reynolds (lubrication); the pressure drop along the fracture depends on the viscosity of the fluid, and the permeability of the rock (leading to fluid 'leak-off'); the fracture aperture depends on the stiffness of the rock mass and the fluid pressure distribution along the crack; and fracture extension depends on the mechanical energy supplied to the region around the crack tip. The tip may propagate ahead of the fluid, leading to a 'lag,'a dry region between the crack tip and fluid front.

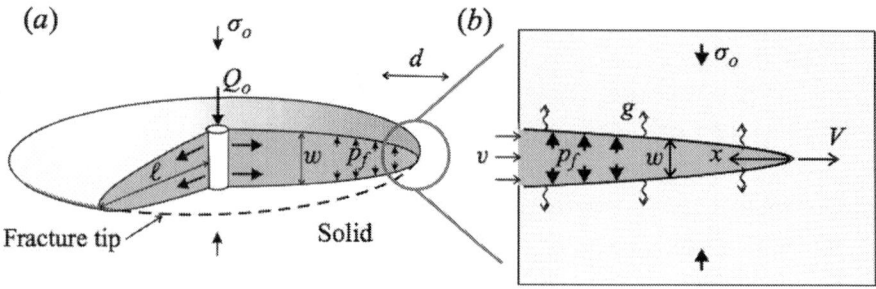

Figure 2: Radial Model of Axi-symmetric Flow and Deformation associated with Hydraulic Fracturing.

Figure 2 illustrates these features for the classical Radial Model in which it is assumed that the fracture propagates symmetrically away from the borehole in a plane normal to the minimum (least compressive) principal in-situ stress, 0.

Development of efficient and robust Hydraulic Fracturing (HF) simulators is central to successful practical HF treatment of petroleum reservoirs. As noted earlier, competing physical processes are operative during the fracturing operation. This has led to a sustained effort over many years to understand and map the multi-scale nature of the tip asymptotics that arise as a result of these competing physical processes in fluid-driven fracture. These asymptotics solutions are critical to the construction of efficient and robust HF simulators. For example, in an impermeable medium, the viscous energy dissipation associated with driving fluid through the fracture competes with the energy required to break the solid material. Breaking of the bonds corresponds to the familiar asymptotic form of linear elastic fracture mechanics (LEFM), i.e., the opening in the tip region is of the form, e.g., (Rice, 1968), with denoting the distance from the tip. However, under conditions where viscous dissipation dominates, the coupling between the fluid flow and solid deformation leads to (Spence and Sharp, 1985; Lister, 1990; Desroches et al., 1994), on a scale that is considerably larger than the size of the LEFM-dominated region, but still small relative to the overall fracture size. In other words, in the viscosity-dominated regime, the zone governed by the LEFM asymptote is negligibly small compared to the crack length. Thus, in the viscosity-dominated regime, the HF simulator should embed a 2/3 power law asymptote rather than the classic 1/2 asymptote of LEFM. Garagash et al.(2011) discuss the generalized asymptotics near the tip an advancing hydraulic fracture, an extension of two particular asymptotics obtained at Schlumberger Cambridge Research Laboratory in the early 1990's (Desroches et al., 1994; Lenoach, 1995).

Three classes of numerical algorithms for HF simulators have now been built: (i) a moving grid for KGD, radial, PKN and P3D fracture simulators; (ii) a fixed grid for plane strain and axisymmetric HF with allowance for a lag between the fluid front and the crack tip, and fracture curving (a versatile code has been developed at CSIRO[9] - Melbourne to simulate the interaction of a hydraulic fracture with other discontinuities); and (iii) fixed grid for simulating a arbitrary shape planar fracture in a homogenous elastic rock. These codes rely on the displacement discontinuity method (Crouch and Starfield, 1983) for solving the elastic component of the problem, i.e., the relationship between the fracture aperture and the fluid pressure.

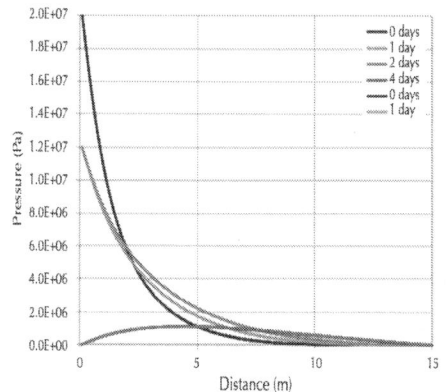 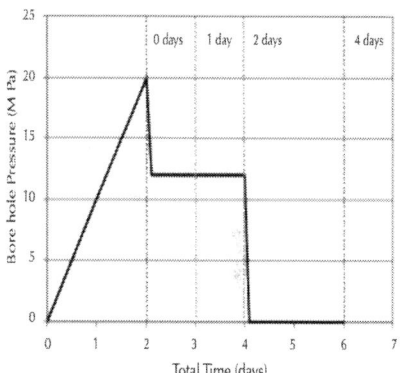

Figure 3: Fluid Pressure Distribution along the Central Axis (Ox) of Figure 1 for a permeable rock due to pressurization and de-pressurization of the borehole.

Figure 3 is presented to illustrate that the fluid pressure in a permeable rock can continue to flow away from the point of injection even after the borehole pressure is reduced to zero. The example shows the distribution of fluid pressure in the rock mass (permeability 5 mD) after (i) 2 days of pressurization up to the peak pressure of 20 MPa in the fracture; (ii) stop pumping and reduce fluid pressure quickly to 12MPa at the point of injection; (iii) hold the pressure constant for 2 days; and (iv) drop the pressure to zero.

It is seen that the pressure in the rock (red curve) has a maximum at some distance from the borehole such that fluid continues to flow into the rock for some time after the pressure in the borehole is reduced to zero. Different combinations of rock permeability, pumping rates and durations can lead to higher peak pressure values in the rock, and longer periods during which fluid can continue to flow away from the well. Such flow may contribute to slip on pre-existing fractures after the pressure in the borehole is reduced to zero.

HYDROSHEAR

Hydraulic fracturing is considered to be initiated from a packed–off interval borehole when the net state of stress around the well bore reaches the tensile strength of the rock. It is important to recognize that fluid pressurization of a well in permeable rock will result in flow of the fluid into the rock as soon as the fluid pressure stimulation process is started. This changes the effective stress state in the rock mass and can lead to slip on pre-existing fractures at fluid pressures below the pressure required to crate and extend a hydraulic fracture. This process of inducing slip on pre-existing fractures is termed 'Hydro-shear'. Flow of pressurized fluid into the rock reduces the effective normal stress ($\sigma n - p$) everywhere in the rock { σn = normal stress at any point; p = fluid pressure.] If c and μ respectively represent the cohesion and coefficient of friction acting across the surfaces of a fracture in the rock, then the effective resistance of the fracture to (shear) sliding, τr, will be:

$$\tau r = c + \mu (\sigma n - p) \qquad (1)$$

Thus, if the pressure p is raised progressively then τr will be reduced correspondingly until it reaches the limit at which sliding will occur. The situation is illustrated graphically in Figure 3. The rock is subjected to a three-dimensional state of stress represented by the principal stresses $\sigma 1$, $\sigma 2$, $\sigma 3$ and the fluid pressure p. The series of points 'X' indicate the effective state of stress on an array of pre-existing fractures in the rock. As illustrated in Figure 5, the effect of increasing the fluid pressure in the medium is to move the stress state on these cracks close to the limiting shear resistance, i.e., to the limiting value represented by the Mohr-Coulomb limit. As the stress state reaches this limit, the cracks will slip. In order to initiate a hydraulic fracture, the fluid pressure would need to be increased further, until the limiting Mohr circle reaches the tensile strength limit of the failure envelope. Since crack surfaces are often not smooth, shear slip will tend to result in crack dilation, and an associated increase in fluid conductivity. It is suggested that hydro-shearing could be more effective than hydraulic fracturing as a stimulation technique in certain applications, e.g., in stimulation of high-temperature geothermal reservoirs. Cladouhos et al. (2011) discuss the application of hydro-shearing as a geothermal stimulation technique. The possibility that silica proppant may dissolve in the aggressive high-temperature fluid environment of some geothermal reservoirs whereas slip on rough fractures develops aperture increase without the need for proppant is also presented as an argument in favor of hydroshearing.

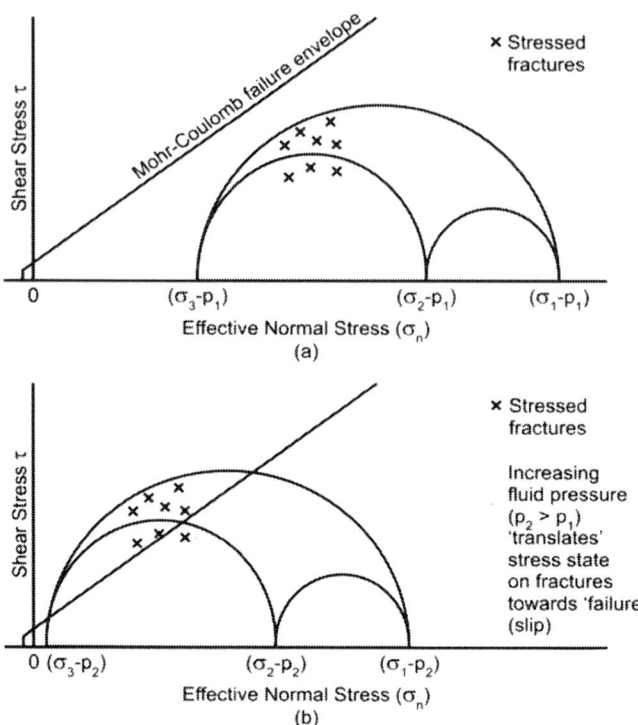

Figure 4: Hydro-shearing — a procedure to generate slip on pre-existing fractures by increasing the fluid pressure to a level below that required to generate a hydraulic fracture.

DEFORMATION AND FAILURE OF ROCK IN SITU

As with fabricated materials, the deformation and failure of brittle rock is also dependent strongly on fractures and discontinuities. In a rock mass, however, the fractures occur over a very wide range of scales from sub-microscopic to the size of tectonic plates. A large specimen of rock will probably include some large fractures, and as the scale of the rock mass increases, fractures from different tectonic epochs.

Study of fracture systems underground in mines and in civil engineering projects allow systems of fractures to be identified and classified statistically into discrete fracture networks (DFN's). The network will include intersecting sets of planar fractures, but individual fractures will tend to be of different lengths, and though organized in two or three spatial orientations, of variable, finite length and not collinear.

Figure 7 presents a two-dimensional illustration of the application of DFN's to the numerical modeling of a fractured rock mass. The in-situ rock mass is considered as a large specimen of intact rock that has been transected by the DFN determined from field observations and fracture mapping underground or at surface outcrops. The properties of the intact rock are built into a Bonded Particle Model of the rock (using the Particle Flow Code (PFC) code) based on results of laboratory tests of the intact rock deformability and strength. The intact rock representation is shown on the left of Figure 6. The DFN (shown on the upper right in Figure 6) then is superimposed onto the intact rock.

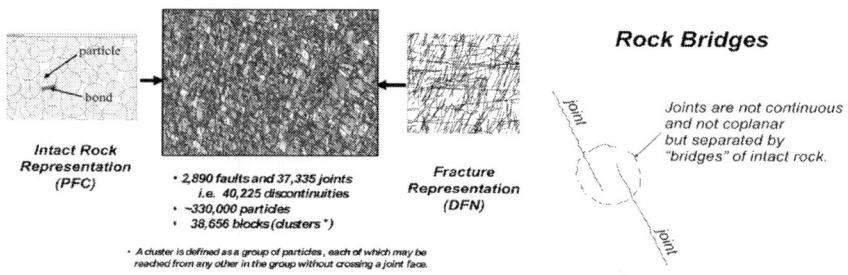

Figure 5: The Synthetic Rock Mass (SRM) representation of a fractured rock mass (in two dimensions).Damjanac et al. (2013) present a discussion of the 'construction' of an SRM in three dimensions.Pierce (2011) presents a comprehensive discussion of practical guidelines and factors involved in the construction of DFN's.

Cohesion and friction values are assigned to the joint planes. [10] - The 'unconfined' strength of a typical large SRM is of the order of a few percent of an intact rock specimen of the same rock

(Cundall, 2008). Much of the in-situ strength is derived, of course, from the in-situ stresses imposed on the SRM in situ. One of the consequences of the finite length and lack of collinearity of joint sets in DFN's is the formation of bridges of intact rock Figure 4 within the SRM. These bridges provide regions of intact rock, and of stress concentration, in the SRM and account for a significant part of the overall strength of the rock mass. Earlier models of a rock mass, considered to consist of several sets of through-going fractures, exhibited much lower rock mass strength (Hoek and Brown, 1980).

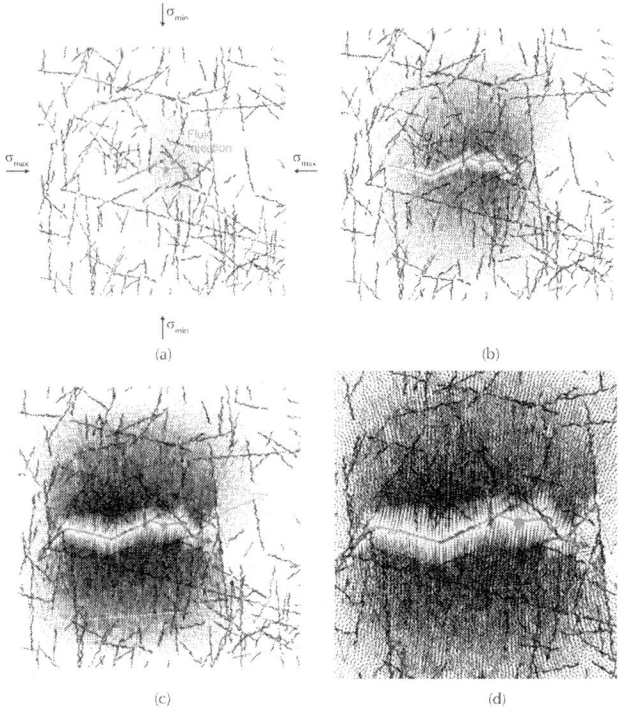

Figure 6: Extracts from simulation of the propagation of a hydraulic fracture in a two-dimensional impermeable SRM (Synthetic Rock Mass). (The horizontal stress σ_{max} is 29 MPa and the vertical stress σ_{min} is 12 MPa – Figure 5(a)). Note that the intact rock between the fractures has a finite strength and can break by rupture of the ce-

mented bonded particles shown in Figure 5. The pressure required to propagate the fracture after breakdown was approximately 10 MPa above the minimum (i.e., least compressive) principal.

Figure 5 presents selected extracts from a two–dimensional *PFC* simulation of the development of a hydraulic fracture in a jointed Synthetic Rock Mass. The SRM model was developed following the procedure outlined in Figure 5. The joint distribution was based on a DFN obtained at the Northparkes Mine in Australia.[11] - Figure 5(a) shows the location of a vertical borehole that was pressurized by fluid until a hydraulic fracture was initiated. The rock mass is assumed to be impermeable. (The path of the fracture has been traced in blue for clarity.) Displacements in the rock mass produced by the hydraulic fracture are shown as vectors on each side of the fracture. It is seen that the fracture started more or less symmetrically on each side of the borehole, but propagation of the right wing was arrested when the hydraulic fracture encountered an adversely oriented pre-existing joint (Figure 5(b)). With increasing pressure, in the borehole, the hydraulic fracture continued to extend asymmetrically towards the left (Figures 5(c) and 5(d) Figure 5(d) is simply an enlarged view of Figure 5(c)). It is seen that the propagating fracture extended partially by opening existing fractures and partially by developing new fractures through intact rock. Although local deviations occur, the overall path of fracture growth is approximately perpendicular to the direction of the minimum compression stress. The existing fractures introduce an asymmetry to the rock mass. In terms of the idealized symmetric crack of Figure 2, the system in Figure 3 can be considered as two cracks, one extending to the right and one to the left of the borehole with a higher 'fracture toughness' on the right compared to the left, etc.

Jeffrey et al. (2009) conducted an underground test in the Northparkes Mine, Australia to observe the propagation of a hydraulic fracture in naturally fractured tock. Figure 7 shows part of the path of the fracture, as seen in a tunnel excavated into the fractured rock. The fracture path shows similar characteristics to those shown in the *PFC* simulation in Figure 6.

Figure 7: Hydraulic fracture (green plastic) crossing a shear zone on the face of a tunnel excavated through the fracture. "The arrows indicate the trace of the fracture with green plastic contained in it. There is no clear fracture between points 1 and 2 but the fracture may have crossed this zone either deeper into the rock or in the rock that has been excavated. Approximately 2 m of fracture extent is visible" (Jeffrey et al., 2009).

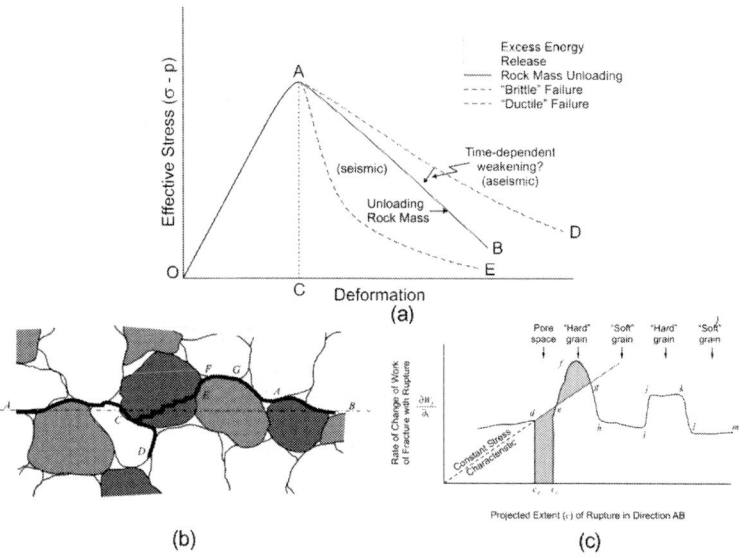

Figure 8: Energy Changes during Propagation of a Fracture through Heterogeneous Rock.

The energy required to initiate crack propagation is represented by the area OAC in Figure 7(a). Whether or not the crack will extend depends on the energy that becomes available from the intact rock around the crack. If the energy released from the rock mass, represented by the area under the red curve AB, is greater than the energy required to extend the crack, represented by the area under curve AE, then the crack will extend; the excess energy represented by the shaded area serves to accelerate the crack and release seismic energy. If the energy required to extend the crack is represented by the area under the green curve AD, it is greater than the energy that would be released from the rock mass, and hence the crack would not extend. It is possible that the crack could exhibit some form of time-dependent weakening (e.g., due to fluid flow to the crack, viscous behavior, etc.) such that the energy required to extend the crack would be reduced. This could lead to crack extension, i.e., as the slope AD increased to overlap AB, but with no excess energy to produce seismicity. Figures 7(b) and 7(c) [12] -illustrate another feature of crack extension on the granular scale. The energy required to extend a crack through or around a grain will be variable; the fracture may encounter pore spaces where no crack energy is required. Application of a constant load to such a heterogeneous system will result in local acceleration and deceleration of the crack-producing bursts of microseismicity. Similar effects can arise in rock fracture propagation at all scales.

It is worth noting that all of these processes of fracture propagation, albeit complex, develop in accordance with the principle of seeking the minimum potential energy of the system.

Much of the preceding discussion has focused on two-dimensional analysis or models. In reality, we are dealing with three- dimensional space (as noted in Figure 6), plus the influence of time (e.g., with respect to fluid flow, or time-dependent rock properties). Figure 8 provides an example from an actual record of hydraulic fracture propagation.

Figure 8 shows the sequence of microseismic events observed during hydraulic fracture stimulation ('treatment' in Figure 8(a)) of a borehole. Early time events are shown as green dots; later events are in red. The microseismic pattern indicates that fracturing started on both sides of the borehole at the injection horizon, but then moved up some 100 m to a higher horizon. As pumping continued, fracturing continued (red locations) on both horizons. It was concluded that the initial fracture in the lower horizon had intercepted a high-angle fault, allowing injection fluid to move to the higher level where it opened up and extended another fracture. Continued pumping led to fracture extension on both horizons. Numerical analysis Figure 8(b) indicated that initial fracture propagation at the lower level resulted in induced tension on the fault above the horizon, but compression on the fault below the lower injection horizon. This explains why injection fluid did not penetrate along the fault below the horizon, and provides a good illustration of the benefit of combining numerical analysis with field observation in understanding fracturing processes.

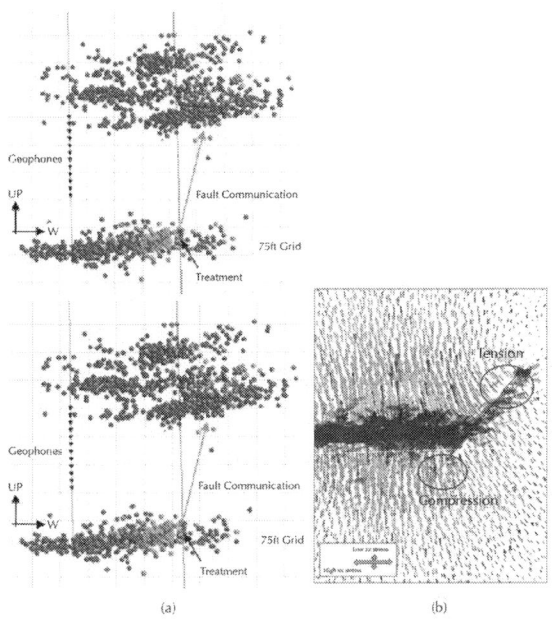

Figure 9: a) Microseismicity observed during hydraulic fracturing in a deep borehole; (b) numerical 'explanation' of the behavior observed in (a).

MICROSEISMICITY AS AN INDICATOR OF SLIP ON FRACTURES

Microseismicity stimulated during hydraulic fracturing and associated stimulation techniques (e.g., hydroshear) is often used to indicate slip and deformation on fractures in the rock. In some cases, it is tacitly assumed that absence of microseismicity indicates absence of slip or deformation. In fact, there is growing evidence that microseismicity does not present a complete picture of deformations induced by stimulation or other effects leading to stress change. Figure 9, reproduced from Cornet (2012) (with permission from the author), shows P-wave velocity changes observed by 4D (time-dependent) tomography during the stimulation of the borehole GPK2 in the year 2000. A detailed discussion of the procedure used to observe and determine the P-wave changes is presented by Calo et al. (2012).

It is seen that the region of detected microseismicity (the cloud of black dots is small compared to the region where the P-wave velocity is reduced by as much as 20% in some regions). Some of the changes in velocity were temporary, suggesting that they may be related to temporal changes in fluid pressure; other changes appeared to be more permanent deformation that occurred aseismically.

These observations indicate that microseismicity, although a valuable indicator of the response of a rock mass to stimulation by fluid injection, does not identify the complete region influenced by a stimulation.

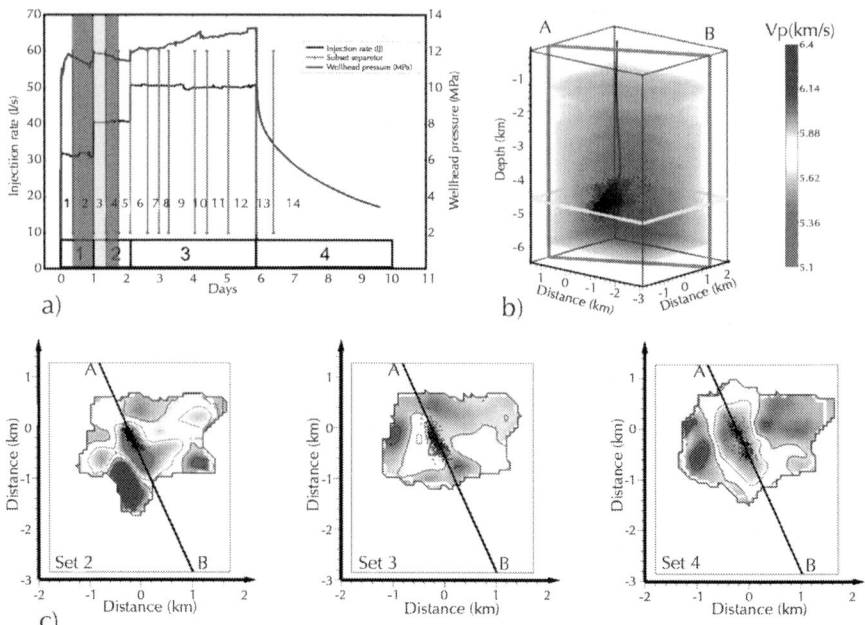

Figure 10: Aseismic slip induced by forced fluid flow as detected by P-wave tomography. (Soultz- sous- Fôrets, France. (a) The injection program (black curve is flow rate, blue curve is well head pressure, horizontal axis is time in days); (b) 3D view of the seismic cloud with respect to the GPK2 borehole. Vertical axis is depth and horizontal axes are distances respectively toward the north and toward the east; and (c) horizontal projections corresponding to the yellow horizontal plane. The vertical green plane is shown as line AB in the plots of part c. P-wave velocity tomography for sets 2, 3 and 4 are indicated respectively by orange, yellow and green colors in the injection program. The vertical axis corresponds to North.

IN-SITU STRESS

As already noted, hydraulic fractures tend to develop in a more or less planar fashion, extending normal to the minimum regional principal stress. Determining the direction, and perhaps the magnitude, of the regional minimum stress is an important element of hydraulic fracturing strategy, especially with the development of directional drilling, which allows borehole to be drilled in the direction considered most favorable for fracturing with respect to stress direction. (see e.g., Figure 15 and related discussion).

Determination of the in-situ stress state also can be a significant challenge.

Stress in rock is distributed throughout the mass, and is influenced by the complicated structure of the mass[13] - . Most techniques of stress determination rely on what are essentially 'point' determinations. One difficulty of determining the regional stress is illustrated by the simple, albeit somewhat artificial, example of Figure 11. This shows a two-dimensional numerical model of the stress distribution in an elastic plate containing several finite frictional fractures.

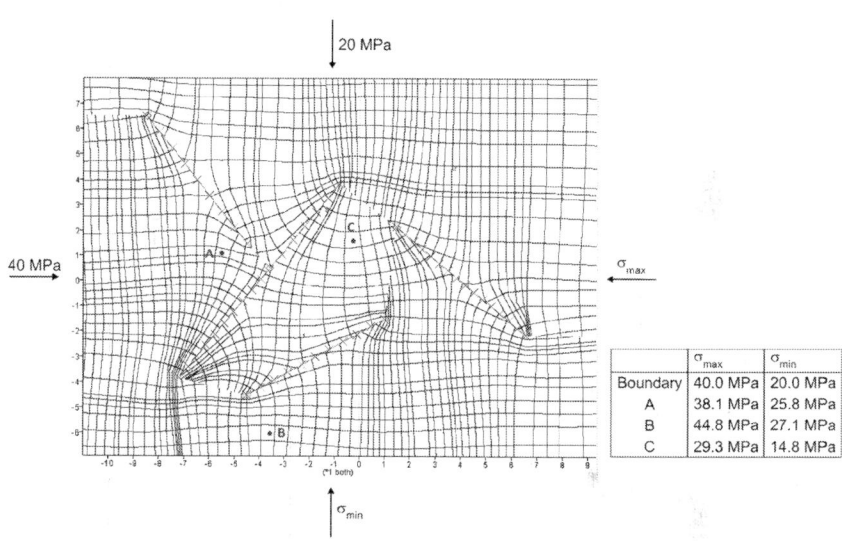

	σ_{max}	σ_{min}
Boundary	40.0 MPa	20.0 MPa
A	38.1 MPa	25.8 MPa
B	44.8 MPa	27.1 MPa
C	29.3 MPa	14.8 MPa

Figure 11: Influence of frictional cracks on the distribution and orientation of principal stresses, illustrative example.

The exercise serves to illustrate the difficulty of making stress determinations from local point measurements, be they in a borehole or on the surface. Stresses can change in orientation and magnitude locally due to geological inhomogeneities, fractures, faults, etc., many of which may be hidden or cannot be observed from the measurement location. Although determinations made at points A and B are reasonably close to the boundary values, point C is considerably different, and the directions of principal stress, as indicated by the principal stress trajectories, can be very different from the (regional) orientations, i.e., at the model boundary.

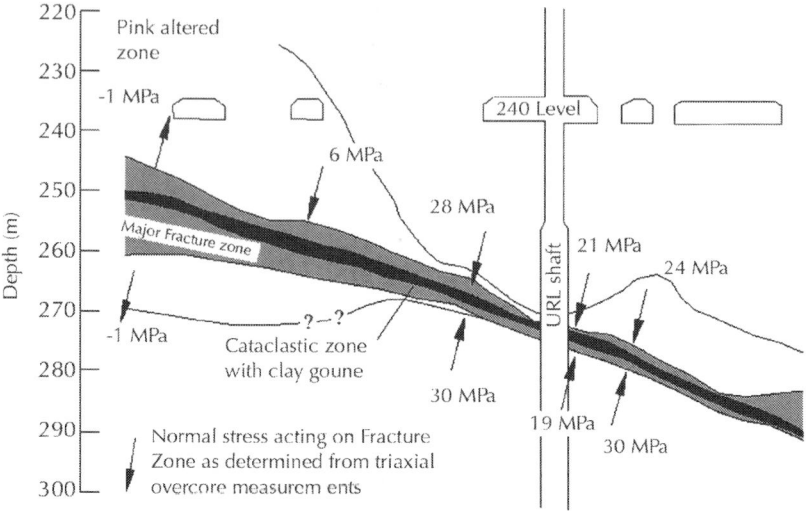

Observed Variability of Normal Stress Across a Thrust Fault Underground Research Laboratory Pinawa, Canada.

Figure 12: Normal stress variation across a thrust fault, Underground Research Laboratory, Canada.

Figure 12 provides an actual example of the variability of stress over relatively short distances. (The vertical and horizontal scales are equal in Figure 12). In this case, the main interest was to assess how normal stresses were affected by the thickness of gouge in the plane of the thrust fault.

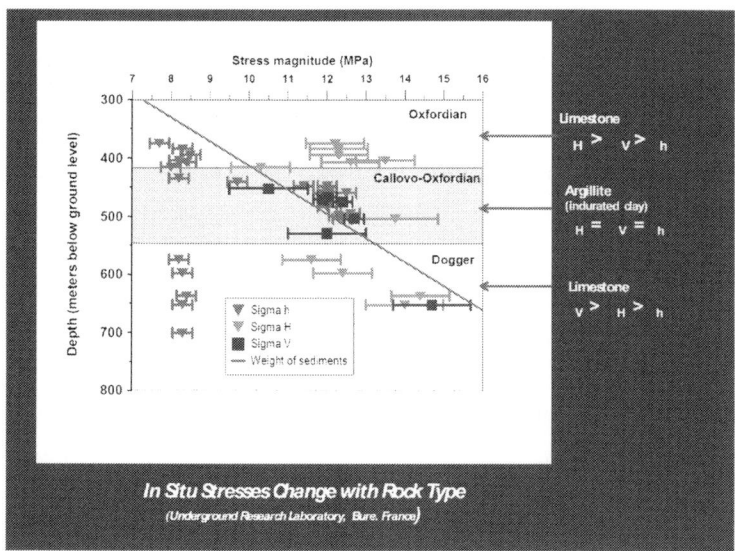

Figure 13: Observed stress distributions in argillite and limestones at the Underground Research Laboratory, Bure, France.

Figure 13 illustrates another important geological influence on stress distribution, changing lithology. This example is from the French Underground Research Laboratory (URL) [14] - at Bure in NE France. Laboratory tests on specimens of the Callovo-Oxfordien Argillite indicate a long-term viscosity of this rock suggesting that any imposed deviatoric stresses would tend towards an isotropic stress state over the order of 10 million years.

Test specimens from the limestones above and below the argillite do not appear to exhibit such viscosity. The stress distributions determined from field measurements support such differences in rheological characteristics of the rock formations.

Commenting on the in-situ stresses observations at Bure (i.e., as shown in Figure 13) Cornet (2012) notes as follows:

"Further, the complete absence of microseismicity in the Paris Basin (Grünthal and Wahlström, 2003,Fig. 4) and the absence of large scale horizontal motion as detected by GPS monitoring (Nocquet and Calais, 2004) indicate that no significant horizontal large-scale active deformation process exists today in this area.

"The important conclusion here is that the natural stress field measured on a 100 km² area at depth ranging between 300 m and 700 m does not vary linearly with depth and is not controlled by friction on preexisting well- oriented faults. Rather, the stress magnitudes seem to be controlled by the creeping characteristics of the various layers rather than by their elastic characteristics, with a loading mechanism that remains to be identified but which is neither related directly to gravity nor apparently to present tectonics.

"It is concluded here that the smoothing out of stress variations with depth into linear trends may be convenient for gross extrapolation to greater depth. But it should not be taken as a demonstration that vertical stress profiles in sedimentary rocks are governed by friction along optimally oriented faults, given the absence of both microseismicity and actively creeping fault. It should not be used for integrating together stress tensor components obtained within layers with different rheological characteristics."

Other examples could be cited, but the message is clear. Determination of in-situ stress in rock is an extremely challenging task, with results subject to considerable variability and uncertainty.

Stress orientations can be estimated from consideration of regional tectonics, faulting and interpretation of evidence from local structural geology supported in some cases by evidence based on borehole logs (e.g., tensile fractures induced along the well bore). Stress magnitudes are, in general, more difficult to determine and usually less significant, except as indicators of how stresses may be distributed across a site where the geology and engineering design are complex. In such cases, interpretation of stress distribution is best done in conjunction with a numerical model of the site, preferably one that includes the influence of important uncertainties and discussion with structural geologists familiar with the area under study.

CRITICAL STRESS STATE' IN THE EARTH'S CRUST

It is sometimes asserted that the Earth's crust is everywhere close to a 'critical state of stress,' i.e., that a small change in the devatoric stress in the rock is likely to produce slip on one or more faults with associated seismic activity. The current global interest in development of major resources of natural gas, the central role of hydraulic fracturing in this development, and the public apprehension that hydraulic fracturing will 'trigger earthquakes' has led to strong opposition to fracturing, and even legislation to ban the use of hydraulic fracturing in some countries and some States in the USA.

As illustrated by Figure 14, the seismic hazard, (i.e., probability of a damaging earthquake) varies very considerably from place to place. Thus, an earthquake of a given magnitude is 1000 times more likely to occur in Southern California than it is in the Eastern United States. The hazard is even lower in regions such as Texas, North Dakota and in the stable Canadian Shield region of the North American tectonic plate. While many earthquakes are initiated at depths considerably greater than depths where hydraulic fracturing is applied, it seems plausible to suggest that there may be less potential for fracturing to induce seismic activity in regions that have low seismic hazard. Also, as indicated by the comments of Cornet in the previous section of this paper, there is evidence that the critical stress hypothesis warrants detailed scrutiny, at least. This could have major implications for development of the world's major natural gas and EGS (enhanced geothermal systems) resources. Two recent studies, National Research Council (2012) and Royal Society – Royal Academy of Engineering (2012), have each concluded that the risk that hydraulic fracturing as used in development of energy resources would trigger significant seismic activity is small, but it would be valuable to examine the critical stress hypothesis more rigorously than has been done to date.

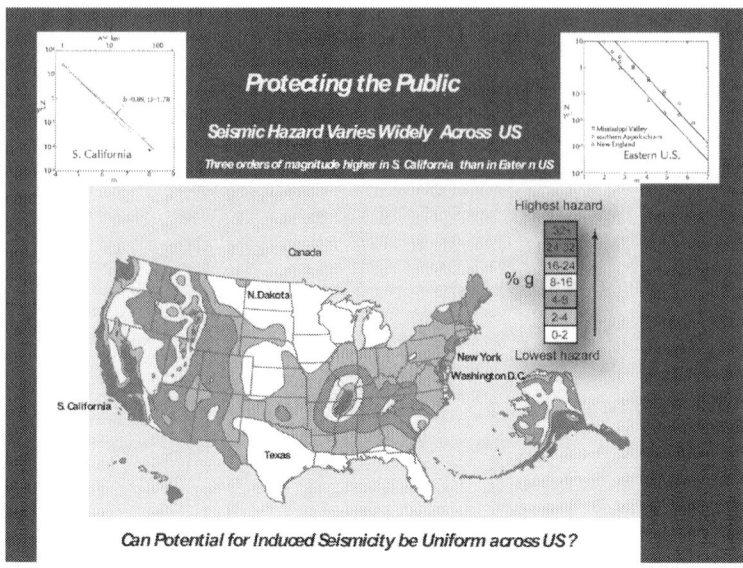

Figure 14: Seismic hazard map of the United States — US Geological Survey.

HYDRAULIC FRACTURING IN TIGHT SHALES

The development of inclined and horizontal drilling (see Appendix 1 - Figure A1-2) has helped stimulate intense activity to develop natural gas production from so-called tight shale, i.e., rock in which natural gas is held tightly within the very fine pore structure of the rock. Figure 15 illustrates the procedure used to stimulate these shales. The well is drilled horizontally in the gas-bearing formation, more or less in the direction of the minimum principal stress. Hydraulic fractures are generated (and propped) at intervals along the well to generate a network of connected flow paths that will allow the gas to flow to the well. Depth (i.e., extent) and spacing of the fractures should be optimized to produce the formations effectively. Bunger et al. (2012) discuss the factors in the design of an effective fracture strategy.

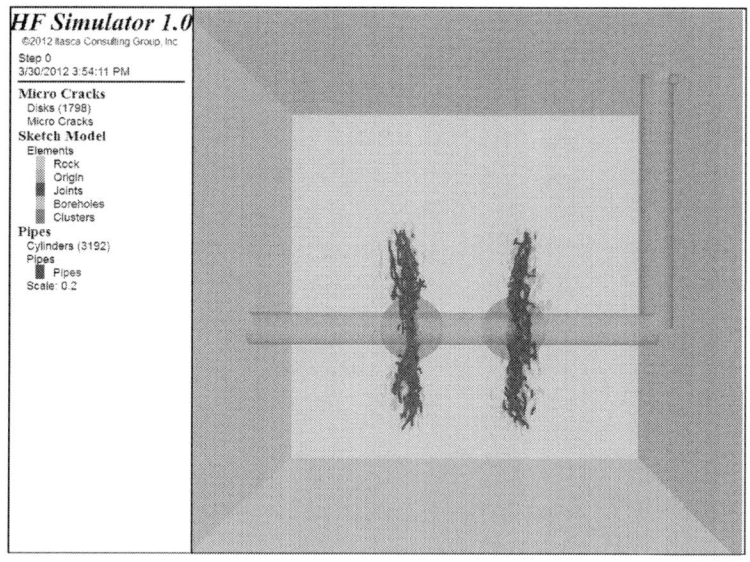

Figure 15: Staged hydraulic fracturing in a horizontal well. There may be many such wells along the horizontal well.

 ## Why Doesn't Microseismicity Correlate With Production?

The Total Rock Volume Affected by Microseismicity Accounts for less Than 1% Gas Production in First 6 Months

Figure 16: The volume of rock defined by micro seismicity is a very small fraction of the volume producing gas.

Figure 16 shows a slide from a recent presentation by Prof. Mark Zoback, who kindly agreed to allow the author to include it here. Although on a somewhat smaller scale, the fact that considerable deformation and fracturing must be taking place that is not associated with detected microseismicity is similar to the phenomena discussed in connection with Figure 10. Prof. Zoback refers to such aseismic deformation as slow slip, and is conducting research to understand the underlying mechanisms, including the possible influence of the clay content of the shale. As can be seen in Figure 17 (courtesy of Prof. Zoback), the clay content can be large.

 ## Average Shale Properties

	BARNETT	MARCELLUS	EAGLE FORD	FLOYD
Depth (ft)	3 - 9,000	2 - 9,500	4 - 13,500	6 - 13,000
TOC (%)	1 - 10	1 - 15	2 - 7	1 - 7
RO (%)	0.7 - 2.3	0.5 - 4+	0.5 - 1.7	0.7 - 2+
Porosity (%)	2 - 14	2 - 15	6 - 14	1 - 12
Qtz + Calcite (%)	40 - 50	40 - 60	50 - 80	20 - 30
Clay (%)	20 - 40	30 - 50	15 - 35	45 - 65
Areal Extent (mi²)	22,000	60,000	15,000	6,000
Resource Size (Tcf)	25 - 250	50 - 500	10 - 100	<<1

How many Floyd Shales are There?

Figure 17: clay content of some typical 'tight' gas shakes.

Figure 18 illustrates the very fine, micron scale, pore structure of a typical tight shale. Although the mechanism(s) by which flow pathways are established in such a fine structure is not clear, the level of microseismic energy release associated with brittle breakage of one or a few bonds will be very small and of high frequency (such that the radiated energy would be rapidly attenuated), and hence, not detectable by any geophone. Thus, absence of microseismicity may not indicate an absence of breakage of brittle bonds. Some mechanism must be operative that generates flow pathways. Intuitively, it might be expected that the clay content of the shale might lead to ductile and viscous deformation that could tend to close the pathways.

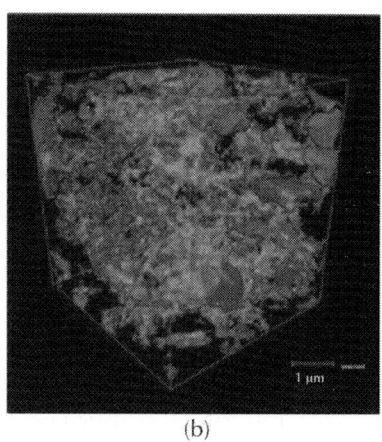

(a)   (b)

Figure 18: a) Outer surface of a FIB-SEM (Focused Ion Beam- Scanning Electron Microscope) volume of Eagle Ford Shale; (b) Transparency view of the distribution of connected pores (blue), isolated pores (red) and organic matter (green). (Courtesy of Prof. Amos Nur and J. Wallis (see Wallis et al., (2012) for details of technology.).

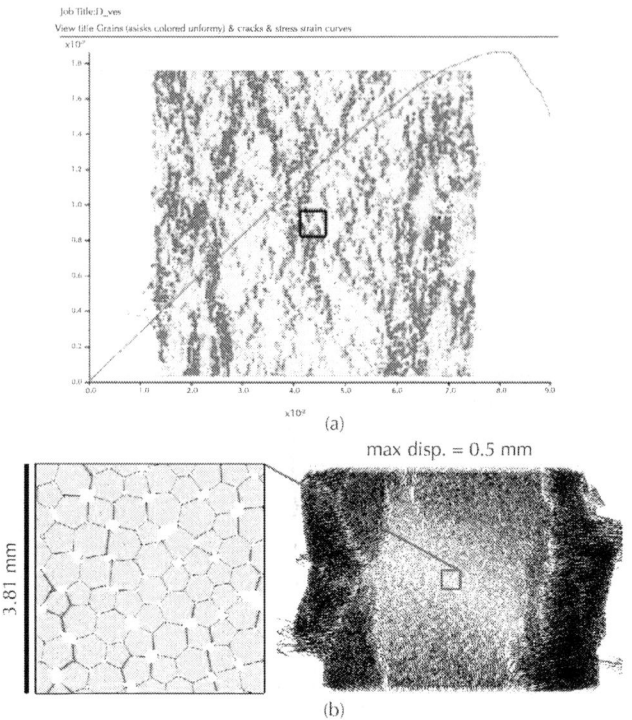

(a)

(b)

FIGURE 19: Micro-rupture of bonds within a *PFC* model of a rock loaded to failure, and beyond, in uniaxial compression. The darker red regions in (a) indicate coalescence of smaller groups of bonds that have ruptured. Eventually these larger regions develop to provide a mechanism that leads to collapse of the specimen. It is seen that bond breakage occurs throughout the specimen as the load is increased. The larger dark red regions will release larger amplitude, lower frequency waves that can be detected, whereas the smaller 'pathways' cannot be detected seismically. The load-deformation curve is shown as an 'overlay' on the specimen.

FRACTURE NETWORK ENGINEERING

This paper has emphasized the central role of fractures in rock, primarily natural fractures developed on a wide spectrum of scales over many tectonic epochs and many millions of years. These fractures and fracture systems are of special significance with respect to hydraulic fracturing and related techniques of fluid injection into rock since the fluid will tend to seek out those fractures that can be more readily opened against the local in-situ stress field as the fluid is injected. Given the complexity and lack of information on the fracture system, stress environment, etc., how can the engineering of hydraulic fracturing and related fluid injection programs advance most effectively?

Confronted with the same complexity of rock in situ, civil engineers and mining engineers have tended to adopt the 'Observational Approach' (Peck, 1969). In essence, this approach involves developing an initial engineering design for the problem, based on a first assessment/estimate of the rock (or soil) properties. Observe the actual performance and modify the initial design as needed to arrive at the desired performance. An example of the Observational Approach (as used in the New Austrian Tunnelling Method) is discussed in Fairhurst and Carranza-Torres (2002), see pp. 24-30. Application of the Observational Approach to Hydraulic Fracturing and related fluid injection techniques faces some disadvantages and some advantages. We do not have 3D access to the engineering site. We do have powerful numerical modeling tools to help make a more informed initial estimate of how the system will perform; and we have sensing systems, both downhole and remote.Figure 20 illustrates a procedure that tries to apply the Observational Approach to hydraulic fracturing and related systems. The illustration describes an application to the extraction of Geothermal Energy.

Stones have begun to speak, because an ear is there to hear them, ...

Cloos, Conversations with the Earth (1954), 4

Microseismicity - predicted and observed

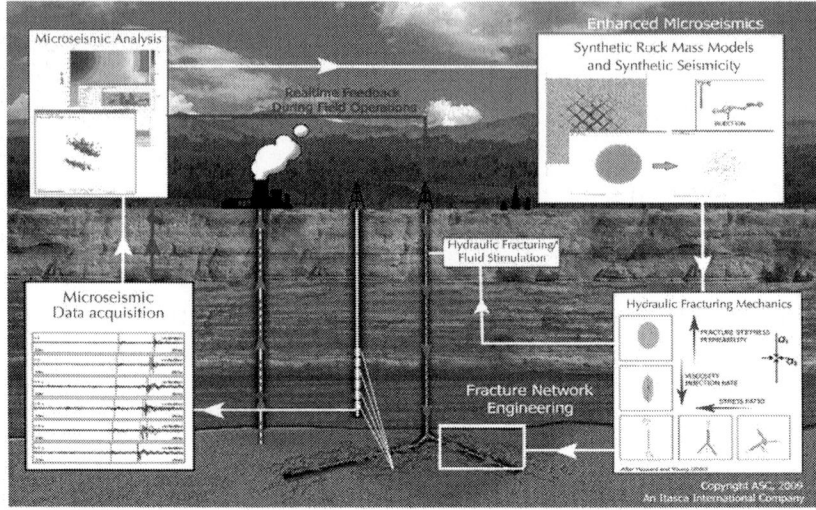

Fracture Network Engineering. Synthetic Rock Mass and Synthetic Seismicity Models are compared with observed microseismic signals for read time control of fracture network development. (Enhanced Geothermal Systems),

Figure 20: Fracture network engineering system.

In this application, an initial design approach is developed based on a numerical modeling study incorporating any available data, insight, etc., on the site. This model provides an initial prediction of the performance. Instrumentation, both downhole and on-surface observes the initial response of the system and compares it with the prediction. This triggers a feedback signal to modify the design input to move the performance closer to the one desired. This iteration continues, changing progressively towards the performance desired.

Although the writer knows of no such Fracture Network Engineering system currently in operation, many of the components are available and it is time to start.

CONCLUSIONS

Expectations for higher living standards of a rising world population, and the associated demand for Earth's resources of energy, minerals and water, lead inevitably to greater focus on resources of the subsurface.

This focus includes the need to develop improved technology to develop these resources, and a better understanding of the nature of the subsurface environment as an engineering material.

Earthquakes and dynamic releases of energy are a daily reminder that on the global scale, Earth is critically stressed, and constantly trying to adjust seeking to achieve a condition of minimum potential energy for the entire system.

On going for many, many millions of years, such adjustments have resulted in the heterogeneous assembly of blocks of rock bounded by essentially planar surfaces; fault, fractures and similar 'discontinuities' varying in scale from tectonic plates and continents down to micron and even nanometers.

Some of these volumes are critically stressed; others are far from a critical condition. National maps of seismic hazards provide evidence of this heterogeneity on a larger scale.

Although Earth Resource Engineering activities may be kilometers in extent, they are small-scale within the larger Earth context. Subsurface engineering in a critically stressed region can be a much different challenge than in a stable region. It is important to assess the initial conditions carefully for each case, and especially where fluid injection is a main component of a project.

The sub-surface is opaque in several ways. Details of the key features that can control the response to an engineering activity in the sub-surface are often unknown. Problems are data-limited. This is particularly the case when the engineering is based on deep borehole systems, as in hydraulic fracturing and related fluid injection technologies.

Although operating in ways that may appear complex, the response of the subsurface to stimulation does obey the laws of Newtonian mechanics, and it is clear that pre-existing natural discontinuities have a major influence on how the subsurface responds to engineered changes.

The advent of powerful computers and developments in numerical modeling provide a potentially major tool to help develop better-informed strategies of subsurface engineering. Used interactively in close conjunction with instrumentation, both downhole and surface based, it should be possible to progressively develop a mechanics-informed understanding and path forward for more effective subsurface engineering.

Much as the field of Fracture Mechanics has led, and continues to lead, to major technological improvements for fabricated materials, so can development of the field of Rock Fracture Mechanics be of transformative value to subsurface engineering, and to society in general.

Hydraulic fracturing and related injection-stimulation systems will certainly be a central element in the future of Earth Resource Engineering. The organizers of HF 2013 are to be commended for focusing attention on this critically important topic.

ACKNOWLEDGEMENTS

Much of the material and concepts discussed in this paper is the result of work and discussions over many years with colleagues at Itasca Consulting Group, Inc. in Minneapolis and faculty in GeoEngineering at the University of Minnesota, especially in this instance, Professor Emmanuel Detournay. Particular help was received from Itasca colleagues Varun, Branko Damjanac, David Potyondy and Mark Lorig, The influence of numerous stimulating discussions with Professor François Cornet of the Institut de Physique du Globe, Strasbourg, France are clearly evident in the paper. Professors Amos Nur and Mark Zoback, of Stanford University, USA and of Ingrain, Inc., Houston, USA assisted with valuable material, as acknowledged in the text. Dr. Rob Jeffrey and Andrew Bunger of CSIRO, Melbourne, the leaders in arranging HF2013, have provided valuable comments, assistance and understanding throughout. To all, I am very grateful. Such invaluable assistance notwithstanding, I accept full responsibility for the interpretations and views expressed in the paper.

REFERENCE

1.   T. L Anderson, 2005Fracture Mechanics: Fundamentals and Applications. 3rd edition, CRC Press (0-84931-656-1

2.   E. V Artyushkov, 1973Stresses in the Lithosphere Caused by Crustal Thickness

3.   J Inhomogeneities, Geophy.Res. 7832November 10, 1973

4.   A. P Bunger, X Zhang, and R. G Jeffrey, 2012Parameters Affecting the Interaction Among Closely Spaced Hydraulic Fractures" SPE Journal March 2012, 292306

5.   M Calo, C Dorbath, F. H Cornet, and N Cuenot, 2011Large scale aseismic motion identified through 4D P-wave tomography; Geophys. J. Int. 18612951314

6.   P. A Cundall, 2008An Approach to Rock Mass Modelling," in From Rock Mass to Rock Model-CD Workshop Presentations (15 September, 2008)- SHIRMS 2008 (Proc. 1st Southern Hemisphere International Rock Symposium, Perth, Western Australia, September 2008) Y. Potvin et al., Eds. Nedlands, Western Australia: Australian Centre for Geomechanics.

7.   T. T Cladouhos, M Clyne, M Nichols, S Petty, W. L Osborn, and L Nofziger, 2011Newberry Volcano EGS Demonstration Stimulation Modeling" GRC Transactions, 35317322

8.   F. H Cornet, 2012The relationship between seismic and aseismic motions induced by forced fluid injections." Hydrogeology Journal (2012) 20: 1463-1466

9.   F. H Cornet, and T Röckel, 2012Vertical stress profiles and the significance of "stress decoupling". Tectonophysics 5812012193205

10. P. A Cundall, M. E Pierce, and D. Mas Ivars. (2008Quantifying the Size Effect of Rock Mass Strength" in SHIRMS 2008 (op. cit.) 2315

11. B Damjanac, and C Fairhurst, 2010Evidence for a Long-Term Strength Threshold in Crystalline Rock,‖ Rock Mech. Rock Eng., 43, 513-531 (2010).

12. B Damjanac, C Detournay, P. A Cundall, and Varun, (2013Three-Dimensional Numerical Model of Hydraulic Fracturing in Fractured Rock Masses" Proc. HF 2013The International Conference for Effective and Sustainable Hydraulic Fracturing, Brisbane, May 20-22, 2013

13. B Damjanac, and C Fairhurst, Evidence for a Long-Term Strength Threshold in Crystalline Rock,‖ Rock Mech. Rock Eng., 43, 513-531 (2010Duchane, D and D. Brown, (2002) "Hot Dry Rock (HDR) Geothermal Energy Research and Development at Fenton Hill, New Mexico" GHC (Geo-Heat Center) Bulletin, December. 2002 1319

14. C Fairhurst, and C Carranza-torres, 2002Closing the Circle-Some Comments on Design Procedures for Tunnel Supports in Rock," in Proceedings of the University of Minnesota 50th Annual Geotechnical Conference (February 2002), 2184J. F. Labuz and J. G. Bentler, Eds. Minneapolis: University of Minnesota, 2002. [available at www.itascacg.comgo to 'About'and Fairhurst Files]

15. C Fairhurst, 1971Fundamental Considerations Relating to the Strength of Rock. Colloquium on Rock Fracture, Ruhr University, Bochum, Germany, April 1971. (see http://www. itascacg.com/about/ff.php)Revised and published in Report of the Workshop on Extreme Ground Motions at Yucca Mountain, August 23-25, 2004, U.S. Geological Survey, USGS Open-File Report 20061277T. C. Hanks et al., Eds. Reston, Virginia: USGS, 2006.

16. J Geertsma, and F De Klerk, 1969A Rapid Method of Predicting Width and Extent of Hydraulic Induced Fractures. J Pet Technol 211215711581SPE-2458-PA. http://dx.doi. org/10.2118/2458-PA

17. J. F Geyer, and S Nemat-nasser, 1982 Experimental Investigation of Thermally induced Interacting Cracks in Brittle Solids Int. J. Solids and Structures 184349356

18. A. A Griffith, 1921 The Phenomena of Rupture and Flow in Solids Phil. Trans. R. Soc. Lond. A 1921,, 221, 163-198 doi:rsta.1921.0006

19. A. A Griffith, 1924 Theory of Rupture. Proc. First Int. Cong. Applied Mech (eds Bienzo and Burgers). 5563 Delft: Technische Boekhandel and Drukkerij. 1924

20. G Grünthal, and R Wahlström, 2003 An Mw-Based Earthquake Catalogue for Central, Northern and Northwestern Europe using a Hierarchy of Magnitude Conversions. J. Seismol. 7, 507-531 (Available at http://seismohazard.gfzpotsdam.de/projects/en/eq_cat/menue_e"q_cat_e.html)

21. E Hoek, and Z. T Bieniawski, 1966 Fracture Propagation Mechanism in Hard Rock," in Proceedings of the First Congress of the International Society of Rock Mechanics. Lisbon, September-October, 1243249 J. G. Zeitlen, Ed. Lisbon: LNEC.

22. E Hoek, and E. T Brown, 1980 Underground Excavations in Rock." Inst'n of Mining and Metallurgy (London) Revised 1982, 164

23. G. C Howard, and C. R Fast, 1970 Hydraulic Fracturing" SPE Monograph 2. Henry L.Doherty Series 203 pp. SPE 30402

24. C. E Inglis, 1913 Stresses in a Plate Due to the Presence of Cracks and Sharp Corners," Trans. Inst. Naval Arch., London, 55(1), 219141

25. R. G Jeffrey, et al 2009 Measuring Hydraulic Fracture Growth in Naturally Fractured Rock. SPE 124919; SPE Annual Technical Conference and Exhibition, New Orleans, Louisiana, USA, 47October 2009

26. J. F Knott, 1973 Fundamentals of fracture mechanics, Wiley (0-47049-565-0

27. National Research Council 2012 Induced Seismicity Potential in Energy Technologies." Washington, DC: The National

Academies Press, 2012. (300p.) View online at http://www. nap.edu/catalog.php?record_id=13355

28.   J. M Nocquet, and E Calais, 2004Geodetic Measurements of Crustal Deformation in the Western Mediterranean and Europe " Pure Appl. Geophy., 161; 661668

29.   R. P Nordgren, 1972Propagation of a Vertical Hydraulic Fracture. SPE J. 124306314SPE-3009-PA. http://dx.doi. org/10.2118/3009-PA.

30.   R. B Peck, 1969Advantages and limitations of the observational method in applied soil mechanics. Geotechnique, 192171187

31.   T. K Perkins, and L. R Kern, 1961Widths of Hydraulic Fractures. J Pet Technol 139937949SPE-89-PA. http://dx.doi. org/10.2118/89-PA.

32.   W. S Pettitt, J. F Hazzard, B Damjanac, Y Han, M Pierce, T Katsaga, and P. A Cundall, Microseismic Imaging and Hydrofracture Numerical Simulations," in Proceedings, 21st Canadian Rock Mechanics Symposium (Alberta, Canada, May 5-9, 2012

33.   M Pierce, 2011Discrete Fracture Network Simulation" DFN training session LOP (Large Open Pit). [ppt slides available on request. Itasca Consulting Group: www.itascacg.com]

34.   A Riahi, and B Damjanac, 2013Numerical Study of Interaction between Hydraulic Fractures and Discrete Fracture Networks" Proc. HF 2013The International Conference for Effective and Sustainable Hydraulic Fracturing, Brisbane, May 20-22, 2013

35.   Royal Society and Royal Academy of Engineering (2012Junep. "Shale Gas Extraction in the UK: a review of hydraulic fracturing" Issued: June 2012, DES2597] View report online at: royalsociety.org/policy/projects/shale-gas-extraction and raeng.org.uk/shale

36.   O Scotti, and F. H Cornet, 1994In-Situ Evidence for Fluid-Induced Asesismic Slip Events along Fault Zones. Int. J. Rock Mech Min.Sci. &Geomech. Abstr. 314347258Control in Mines (1965). South African Institute of Mining and Metallurgy, Johannesburg, 606p

37.   A. M Starfield, and P. A Cundall, 1988Towards a Methodology for Rock Mechanics Modelling" Int. J. Rock Mech. Min. Sci.& Geomech. Abstr. 25 (3) 99106

38.   J. F Tester, et al2006The Future of Geothermal Energy"-Impact of Enhanced Geothermal

39.   Systems (EGS) on the United States in the 21st CenturyMIT Press.

40.   J. D Wallis, J Devito, and E Diaz, 2012Digital Rock Physics-A New Approach to Shale Reservoir Evaluation" Oilfield Technology,   March   2012   [http://www.ingrainrocks.com/articles/a-new-approach-to-shale-reservoir-evaluation/]

41.   X Zhang, and R. G Jeffrey, 2008Re-initiation or termination of fluid-driven fractures at frictional bedding interfaces" JGR, 113BO 8416, doi:10.1029/2007JB005327,

Recovery of Benthic Megafauna from Anthropogenic Disturbance at a Hydrocarbon Drilling Well (380 M Depth in the Norwegian Sea)

AndrewR. Gates and Daniel O. B. Jones

Ocean Biogeochemistry and Ecosystems Group, National Oceanography Centre, Southampton, United Kingdom

ABSTRACT

Recovery from disturbance in deep water is poorly understood, but as anthropogenic impacts increase in deeper water it is important to quantify the process. Exploratory hydrocarbon drilling causes physical disturbance, smothering the seabed near the well. Video transects obtained by remotely operated vehicles were used to assess the change in invertebrate megafaunal density and diversity caused by drilling a well at 380 m depth in the Norwegian Sea in 2006. Transects were carried out one day before drilling commenced and 27 days, 76 days, and three years later. A background survey, further from the well, was also carried out in 2009. Porifera (45% of observations) and Cnidaria (40%) dominated the megafauna. Porifera accounted for 94% of hard-substratum organisms and cnidarians (Pennatulacea) dominated on the soft sediment (78%). Twenty seven and 76 days after drilling commenced, drill cuttings were visible, extending over 100 m from the well. In this area there were low invertebrate megafaunal densities (0.08 and 0.10 individuals m^{-2}) in comparison to pre-drill conditions (0.21 individuals m^{-2}). Three years later the visible extent of the cuttings had reduced, reaching 60 m from the well. Within this area the megafaunal density (0.05 individuals m^{-2}) was lower than pre-drill and reference transects (0.23 individuals m^{-2}). There was a significant increase in total megafaunal invertebrate densities with both distance from drilling and time since drilling although no significant interaction. Beyond the visible disturbance there were similar megafaunal densities (0.14 individuals m^{-2}) to pre-drilling and background surveys. Species richness, Shannon-Weiner diversity and multivariate techniques showed similar patterns to density. At this site the effects of exploratory drilling on megafaunal invertebrate density and diversity seem confined to the extent of the visible cuttings pile. However, elevated Barium concentration and reduced sediment grain size suggest persistence of disturbance for three years, with unclear consequences for other components of the benthic fauna.

INTRODUCTION

Exploratory hydrocarbon drilling activities are increasing in deeper water [1], [2] and in more environmentally sensitive areas [3]. Environmental impacts associated with offshore exploration drilling include the discharge of cuttings on to the seabed [4], discharge of produced water [5]and the possibility of a major blow out or oil spill [6]. By their nature blow outs and oil spills are unpredictable events, but disturbance from cuttings is well regulated and monitored, providing a useful opportunity to study disturbance in inaccessible and normally quiescent deep waters.

In modern best-practice exploration drilling, disturbance to the seabed at well locations results from the discharge of a mixture of drill cuttings and water-based drilling mud (fluid used to lubricate the drill bit and maintain the structural integrity of the well).This occurs during the initial phase of drilling when the widest diameter sections of the hole are drilled (the "top-hole"), before the marine riser and blow-out preventer (BOP), a large metal structure sitting on top of the well, are deployed. This disturbance is characterized by a combination of physical smothering of the seabed, associated changes in sediment structure, and the potential toxic effects of exposure to the chemical constituents of the mud used in the drilling process [7], [8],[9]. Barite is often added as a weighting agent in drilling mud so barium is a frequently used tracer for drilling disturbance [10], [11]. After deployment of the BOP the cuttings and mud are re-circulated to the surface, cleaned and discarded from the rig. In contrast to this practice, older methods of exploration drilling discharged greater quantities of oil-based drilling mud and cuttings to the seabed.

Exploration drilling disturbance initially results in reduced abundance and diversity of the meiofaunal [12], macrofaunal [13], [14] and megafaunal [4], [15] components of benthic communities. The deposition of cuttings will also affect sediment bacteria, which can comprise up to 90% of benthic biomass [16]. Reduced benthic diversity, in turn, may result in reduced ecosystem functioning [17]. In addition, there is some experimental evidence that drilling disturbance changes overall ecosystem functioning. Biogeochemical fluxes from the sediment (leading to oxygen depletion in the sediment) were altered immediately after addition of cuttings, and bioturbation inhibited by increased sedimentation [8], [18], [19].

In the north-east Atlantic, where water-based drilling mud is used, exploration drilling usually has an impact on the seabed, visible in remote video survey, extending 100 to 200 m from the well. This results in reduced sediment heterogeneity and significant reductions in megafaunal abundance and diversity shortly after the disturbance [4], [20]. According to older studies, which report on disturbance from oil-based drilling mud, hydrocarbon drilling in shallower water leads to altered sediment characteristics with resultant changes to macrobenthic communities over larger areas [11], [20], [21]. Even in more accessible shallower areas it is unclear how long the effects of such disturbance persist [22] and few studies of recovery from any form of anthropogenic disturbance have been carried out in deep water [23], [24].

Recovery typically implies the return of an ecosystem to pre-disturbance conditions as a result of the operation of homeostatic ecological control mechanisms [25]. Recovery is a complex phenomenon involving various spatially and temporally dynamic biotic and abiotic changes. The recovered ecosystem may be altered in some way from its original state, for example in terms of function, structure, species composition or diversity [25].

The benthic megafauna includes those organisms over 1 cm that inhabit the sediment-water interface [26]. Benthic megafaunal organisms play a key role in the functioning of deep-sea ecosystems [27]. Through their actions such as burrowing and feeding they redistribute sediment and influence local scale biogeochemistry [28], [29]. The presence of sessile forms may influence habitat heterogeneity [30]. The megafauna may be affected in several ways by drilling disturbance. For example, physical smothering has been shown to induce increased stress protein expression in motile forms [31] while sessile suspension feeding organisms may also be negatively affected by sedimentation [32].

The well-documented and relatively accessible nature of exploration drilling disturbance provides a valuable opportunity to investigate the process of recovery of benthic megafauna in deeper water. Through the SERPENT project [33] a time-series study of the benthic invertebrate megafauna was carried out around an exploration well at the Morvin field in the Norwegian Sea. Surveys were conducted before drilling, and 27 days, 76 days and three years after drilling and addressed four objectives: 1) to describe the megafaunal species diversity and abundance at the Morvin location, 2) to identify the temporal change in the visible extent of drill cuttings disturbance, 3) to carry out a local-scale, time-series assessment of recovery of benthic megafaunal invertebrates from hydrocarbon drilling disturbance, 4) to use evidence of bioturbation as an indicator of ecosystem function along a disturbance gradient. These objectives are designed to test the hypothesis that over a period of three years physical and biological processes redistribute drill cuttings and water based mud to an extent that megafaunal organism abundance and diversity can recover after an initial physical disturbance from exploration drilling in deeper water.

METHODS

Ethics Statement

No specific permits were required for the described field studies. The site was part of Statoil's production licence 134b and subject to oil drilling activities. No invertebrate megafauna specimens were collected as the work was carried out using video techniques.

Study location

The Morvin field is located on the continental slope of the Norwegian Sea (Figure 1). On 24th March 2006 drilling commenced on an exploration hydrocarbon well from the semi-submersible drilling rig *West Alpha* in 380 m water depth; position 380172 E, 7224481 N. Positional information was recorded in Universal Transverse Mercator (UTM) zone 32 N based on the European Datum 1950 (ED50).

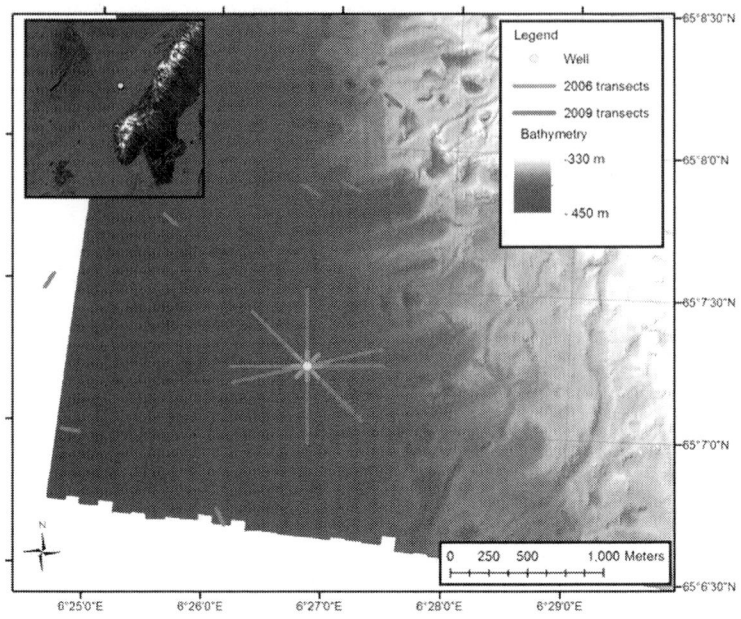

Figure 1: The Morvin survey design.

The 2009 video transect survey is shown in red. Previous surveys were at the same location with 100 m video transects radiating from the well and are shown in green. The location of the Morvin field in the Norwegian Sea is shown as an inset.

Data Collection

Video Surveys

Three video transect surveys were carried out in 2006 using an Oceaneering Hydra Magnum 041 work-class drill-support Remotely Operated Vehicle (ROV) launched directly from the West Alpha. Each survey comprised eight transects, approximately 100 m in length, limited by ROV tether length (owing to launch from the stationary drilling rig). Surveys were carried out one day before (23rd March 2006), 27 days after (21st April 2006) and 76 days (9th

July 2006) after drilling. The straight-line transects radiated from the well location in 8 directions (0, 45, 90, 135, 180, 225, 270, 315°: Figure 1). Transects conducted before drilling followed a set heading (using the ROV gyrocompass) from a buoy marking the intended well position. Distance from the well was estimated from the amount of ROV tether unwound. After drilling ROV sonar was used to improve navigational precision and transects were flown towards the BOP (a clear sonar target). The ROV was flown at a speed of approximately 0.2 m s^{-1} with the standard-definition colour video camera (Kongsberg OE1366) approximately 1 m above the seabed. The camera was positioned at an angle of 18° from horizontal (the maximum angle possible without viewing the ROV frame) with the zoom set to maximum wide angle. Transect width (mean of 1.0 m; max variation±0.2 m) was calculated from the camera acceptance angles and verified following Jones et al. (2006). A digital stills camera (Kongsberg OE14-108) was used to obtain high-resolution photographs of organisms for species identification in separate, opportunistic surveys. The pre-drilling SW transect was omitted from further analysis owing to poor visibility.

Over the 3rd to 4th May 2009, more than three years after drilling commenced, an additional video survey was carried out from the vessel *Acercy Petrel* equipped with the Acergy Solo MKII survey class ROV. Four video transects of 1 km length were carried out, crossing the well at their mid point. For comparison, ten reference transects were also taken (Figure 1). These were 100 m in length, between 1 and 3 km from the well. Starting points and headings for the reference sites were randomly selected. The Morvin area had been the subject of extensive deep-water coral reef mapping and studies of seabed fluid flow [34], [35]; thus any reference transects located near possible reef features were rejected and another random starting point and heading generated.

Recording of the transects began and ended 20 m beyond the planned positions to ensure that the correct altitude and speed were attained before the intended start/finish point. The ROV was flown at approximately 0.3 m s^{-1} with the camera height of approximately 2.5 m above the seabed. The colour video camera (IMENCO Z

1051) was as close to vertical as possible at an angle of 24° below the horizontal with the zoom set to maximum wide angle (mean transect width of 2.6 m; max variation±0.3 m). UTM positional data (from Ultra-Short Baseline Navigation) were continually recorded. The greater ROV altitude in this survey is because of differences in equipment associated with the survey carried out from a ship in contrast to the earlier surveys which were carried out from a drilling rig and may cause some variation in both species density and diversity measurements.

The four surveys described above will be referred to as "Pre" (1 day before drilling), "Post 1" (27 days after drilling commenced), "Post 2" (76 days after drilling) and "Post 3" (three years after drilling). Reference sites studied three years after drilling are referred to as "R" sites.

Additional Data Collection

Sediment samples were collected using ROV push corers before and after drilling. Before drilling, single samples were collected from the well location at approximately 50 and 100 m north of the well. After drilling (21st April 2006) single samples were collected at 10 m and 100 m north east and west of the well. The samples were retrieved to the surface, the depth of an visible drill cuttings measured and the top 50 mm retained and frozen.

Five graduated marker buoys were deployed around the well before drilling commenced. The marker buoys were placed at eight metres north, east and west of the well and at 50 m and 100 m north east of the well. Observations of sediment accumulation around the buoys were made using the ROV at intervals during the drilling programme. The buoys were removed from the seabed at the end of the drilling programme in 2006. In 2009 three replicate sediment samples were collected using ROV push corers at 25 and 50 m from the well on four headings (N, NW & SW). On all headings the samples were divided into 0–20, 20–40 and 40–60 mm sections and preserved by freezing. It was not possible to collect the planned samples at 100 m from the well because of time limitations.

Video Data Analysis

In all cases, video was replayed at half speed and every individual animal was counted and its position recorded as it passed the bottom of the screen. Colonial organisms were counted as single individuals. Megafaunal organisms were identified to the lowest possible taxonomic level. Where species identification was not possible, operational taxonomic units (OTU) were used. Fish were excluded from analysis of benthic abundance data because of their motility and tendency of some species to follow the ROV. Megafaunal density was calculated from abundances divided by the area of the transect section (transect section length multiplied by image width). Features on the seabed such as rocks and burrows were recorded and all data were plotted in a geographic information system using the software ArcGIS (version 9.3).

The distribution of drill cuttings was assessed visually from the video footage. Disturbed sediment was recognized on the basis of its characteristically pale colour and absence of visible evidence of bioturbation (Table S1). The boundaries of the disturbed area were identified and mapped. Megafaunal datasets were extracted from these zones in ArcGIS for comparison of the disturbed zones with other areas.

Data for each well-site transect were split into 100 m distance zones. In the post-drilling surveys part of the 0–100 m zone was visibly disturbed, so this sample unit was split into two sections "Disturbed" and "Beyond Disturbance" in order to identify the effects of disturbance at the highest resolution possible with video observations. For statistical analysis the pre-disturbance transects were split into the same sections as described above (based on the disturbance extent in Post 1 in 2006) so that the densities of fauna in the pre-drilling samples were properly compared in the statistical model. Results were presented based on the disturbance zones rather than consistent distance zones in order to identify the impact after three years.

To describe abundance, both total density and density of organisms associated with different substrata were calculated.

A range of diversity indices were calculated to assess both the evenness and species richness elements of diversity [36]. Sampling units were of variable area so species richness (S) was rarefied to 50 individuals ($ES_{(50)}$). Evenness was calculated as Pielou Evenness (J'). In addition, the widely-used Shannon-Wiener Index (H') was presented to allow comparison with other studies. These measures were calculated using the software package PRIMER v.6 [37].

Three generalized linear (GLM) statistical models were independently developed [38] to examine whether the density (no. m^{-2}) of total, sessile and motile megafauna at Morvin could be explained using the explanatory variables distance and year. Random sites were coded with a distance of >1000 m from drilling and included in all analysis. All explanatory variables were treated as categorical data. The model was fitted with quasi-Poisson errors using the R function GLM and the ANOVA function of the R package CAR (companion to applied regression) [39] in the R programming environment [40].

The megafaunal assemblage composition was investigated using multivariate analyses. A fourth root transformation was applied to buffer the influence of dominant taxa and similarities were calculated using Bray-Curtis coefficients [41]. The similarity values were subjected to both classification (hierarchical group-average clustering) and ordination (non-metric multi-dimensional scaling, MDS) using the PRIMER software. The difference in the megafaunal assemblage composition was assessed using two-way permutational multivariate analysis of variance (PERMANOVA) [42] with distance zones and survey time as factors. PERMANOVA was implemented using the R package Vegan [43].

In addition to the megafauna, structures on the seabed were documented. Rocks were counted and used in later analysis to document the background environment. Conspicuous burrows in the sediment (likely decapods, *Geryon* sp. – Figure S1) were also counted as an indicator of bioturbation activity along the disturbance gradient.

Recovery

Response Y, which represents recovery of the benthic environment after disturbance [44], was calculated based on the percentage change from mean "pre-drill" values of the following indices of diversity: mean motile organism density, sessile organism density, species richness, evenness, Shannon Wiener diversity and Bray-Curtis similarity. Response Y is the percentage difference between impacted and control sites. In order to prevent a right-skewed distribution[44], it is presented transformed as follows (where X is the percentage difference from the pre-drill survey):

$$y = \log e(1 + [x/101])$$

Variation in response Y was tested using two-way ANOVA on ranks with the factors distance and year using the R package.

Environmental Data

Chemical (heavy metals) and particle size distribution analyses were conducted on the sediment samples. Heavy metals analysis (Cd, Pb, As, Se, Sn) was carried out using atomic absorption spectroscopy (Perkin Elmer SIMAA 6000). The method applied was in accordance with Norwegian standard NS4770 and consisted of a partial acidic extraction using 7 NHNO3 in an autoclave. Mercury was analysed according to the same standard but using a different instrument (CETAC M-6000A Hg Analyzer). Thirty other elements were analysed according to the same standard using ICP-AES (Perkin Elmer Optima 4300 Dual View). Particle size distributions were determined using a Coulter LS200 instrument in the range 0.4–2000 μm.

During the surveys depth and temperature were measured using a ROV-mounted sensor (Paroscientific Digiquartz® 8 series).

RESULTS

The Background Environment

The well was located at 380 m depth. There was no appreciable depth variation within 100 m of the well but beyond this, in the 2009 survey area (well site and reference video transects) depth varied between 362 m and 397 m. The predominantly flat sediment was punctuated by small rocks providing some hard substratum. Decapod burrows in the soft sediment were an important feature of the environment. Seabed water temperature was 7.4°C on both 20th April 2006 and 3rd May 2009. Salinity was 35.5 on the same date in 2009, but was not measured in 2006.

The invertebrate megafauna observed during the background quantitative video surveys (Pre-drill and R transects) comprised 27 distinct taxa with a total density of 0.22 m^{-2} (examples shown in Figure 2 and listed in Table 1). Additional taxa were observed across all the disturbance transects. The megafauna was dominated by Porifera (44.5% of total fauna) and Cnidaria (40.6%). The Echinodermata (11.6%) were also important. Of the Cnidaria, soft-sediment dwelling pennatulid octocorals were most abundant and were represented by four distinct taxa, of which *Kophobelemnon stelliferum* was the most common (24.5% of all observations). There were nine distinct poriferan taxa, which were predominantly attached to hard substrates; *Phakellia* sp. (13.5%) and the unidentified "tiny white sponge" (9.2%) were the most abundant. The echinoderms were dominated by the deposit-feeding holothurian *Parastichopus tremulus* (8.3%).

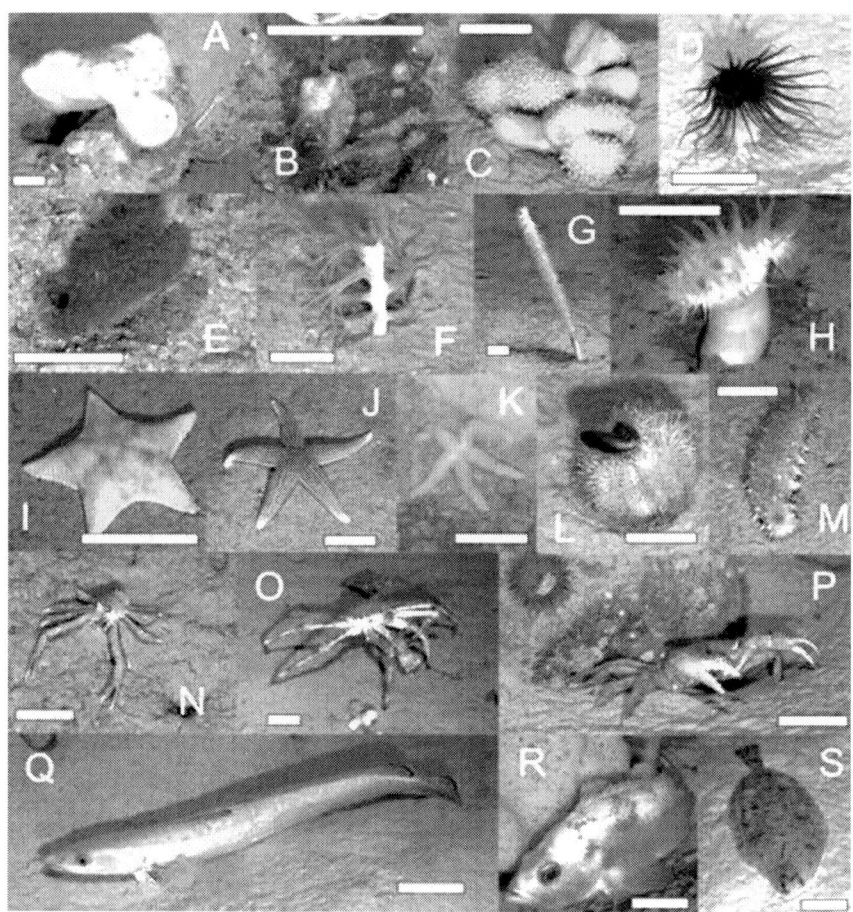

Figure 2: Examples of the megafaunal taxa observed at Morvin.

A: *Mycale* sp., B: *Hymedesmia* sp., C: *Alcyonium* sp., D: *Cerianthus* sp., E: *Pennatula phosphorea*, F: *Kophobelemnon stelliferum*, G: *Funiculina* sp., H: *Bolocera* sp., I:*Porania* sp., J: *Asterias rubens*, K: *Henricia* sp., L: *Echinus* sp., M: *Parastichopus tremulus*, N: *Munida* sp., O: *Lithodes* sp., P: *Geryon* sp., Q: *Molva molva*, R: Sebastidae, S: *Glyptocephalus cynologus*. Scale bar on images represents 50 mm.

Table 1: Mean megafaunal taxon density (per 100 m^{-2}) from video observations before, during and after the drilling operations at increasing distance from the well

Likely species/morphotype	Substratum	Background Pre	R	Post drill 1 Dist	Beyond	Post drill 2 Dist.	Beyond	Post drill 3 Dist.	Beyond	100–200 m	200–300 m	300–400 m	400–500 m
Phakellia sp. 1 †	H	3.18	2.97	0.37	0.00	3.67	1.99	0.16	1.09	1.30	1.32	0.61	0.62
small spherical white sponges	H	2.32	2.04	0.24	0.16	0.09	1.50	0.52	1.20	1.97	1.57	1.48	1.55
Mycale sp.†	H	0.42	2.51	0.12	0.16	0.09	0.99	0.35	0.59	1.01	1.49	0.86	0.84
Haliclona sp.	H	1.23	0.94	0.12	0.37	0.81	0.49	0.16	0.69	0.71	1.01	0.69	0.92
Encrusting white sponge	H	1.60	0.65	0.24	0.00	0.09	0.27	0.47	0.48	0.45	0.53	0.10	0.20
Stylocordyla borealis	S	0.49	0.51	0.00	0.00	0.00	0.12	0.00	0.41	1.01	0.79	0.55	0.68
Hymedesmia sp.†	H	0.42	0.04	0.00	0.00	0.00	0.27	0.00	0.20	0.20	0.20	0.00	0.05
Acinella sp.†	H	0.26	0.36	0.00	0.00	0.09	0.27	0.00	0.00	0.00	0.00	0.00	0.00
Phakellia sp. 2	H	0.00	0.08	0.00	0.00	0.00	0.31	0.00	0.00	0.00	0.00	0.00	0.00
Kophobelemnon stelliferum †	S	5.12	5.63	4.01	6.91	1.93	8.52	0.87	3.99	3.85	3.65	2.92	2.90
small straight pennatulid	S	0.83	1.39	0.18	0.48	1.39	1.80	0.63	1.08	1.25	1.30	1.57	0.90
Funiculina sp.†	S	0.14	1.74	0.00	0.37	0.09	0.63	0.00	0.42	0.51	0.36	0.28	0.78
Cerianthus sp.†	S	0.00	0.39	0.35	0.21	0.19	0.43	0.00	0.53	0.14	0.10	0.29	0.50
Bolocera sp.†	H	0.53	0.31	0.21	0.48	0.20	0.47	0.00	0.24	0.00	0.20	0.15	0.14
Pennatula phosphorea†	S	0.39	0.36	0.12	0.43	0.09	0.52	0.00	0.08	0.15	0.16	0.14	0.00
Alcyonium sp.†	H	0.00	0.36	0.00	0.00	0.09	0.27	0.19	0.08	0.05	0.10	0.05	0.00
red cnidarian	H	0.00	0.00	0.00	0.00	0.00	0.00	0.00	0.00	0.16	0.05	0.05	0.00
Lophelia pertusa	H	0.00	0.04	0.00	0.00	0.00	0.00	0.00	0.00	0.00	0.00	0.05	0.10
Colus sp.	G	0.14	0.08	0.00	0.00	0.00	0.36	0.00	0.00	0.05	0.00	0.00	0.00
Geryon sp.†	S	0.28	0.08	1.31	2.14	0.00	1.30	0.00	0.25	0.05	0.05	0.04	0.00
Pandalus sp.†	G	0.42	0.11	0.00	0.08	0.37	0.88	0.13	0.33	0.35	0.15	0.05	0.14
Lithodes sp.†	G	0.00	0.00	0.00	0.00	1.18	0.35	0.00	0.00	0.05	0.05	0.00	0.00
Munida sp.†	G	0.14	0.04	0.00	0.00	0.00	0.18	0.74	0.12	0.05	0.11	0.05	0.05
Bryozoan	H	0.00	0.00	0.00	0.00	0.00	0.12	0.00	0.16	0.21	0.11	0.10	0.19
Nipponemertes sp.	S	0.00	0.00	0.00	0.00	0.00	0.10	0.00	0.10	0.00	0.05	0.05	0.00
echiuran*	S	0.00	0.00	0.00	0.00	0.00	0.00	0.00	0.00	0.00	0.00	0.00	0.00
Parastichopus tremulus†	S	2.62	1.68	0.00	0.46	0.00	0.52	0.69	1.18	1.80	1.25	1.39	1.82
Henricia sp.†	G	0.00	0.20	0.34	0.00	0.09	0.53	0.27	0.08	0.14	0.15	0.10	0.05
Ceramaster sp.	G	0.26	0.23	0.00	0.00	0.00	0.28	0.00	0.11	0.14	0.21	0.30	0.36
Asterias rubens†	G	0.00	0.00	0.00	0.00	0.00	0.76	0.00	0.00	0.00	0.00	0.00	0.00
Porania sp.†	G	0.00	0.20	0.00	0.00	0.00	0.16	0.00	0.00	0.14	0.00	0.05	0.05
Hippasteria sp.	G	0.00	0.00	0.00	0.00	0.00	0.00	0.00	0.00	0.00	0.10	0.00	0.00
Crossaster sp.*	G	0.00	0.00	0.00	0.00	0.00	0.00	0.00	0.00	0.00	0.00	0.00	0.00
Cidaris cidaris	G	0.14	0.12	0.00	0.00	0.00	0.00	0.00	0.00	0.00	0.00	0.05	0.00
Echinus sp.†	G	0.14	0.08	0.00	0.00	0.09	0.37	0.00	0.12	0.10	0.15	0.26	0.21
Indet 2	H	0.00	0.00	0.00	0.00	0.00	0.00	0.00	0.08	0.05	0.00	0.00	0.00
Total density		21.05	23.11	7.63	12.18	10.57	24.68	5.19	13.52	15.86	15.17	12.25	13.05

* = species that were observed but not recorded, either outside of survey area or large motile organisms intentionally excluded to prevent over estimation of abundance. { = higher resolution still photograph collected, otherwise only recorded from video footage. H = hard-substratum organisms, S = soft-substratum organisms, G = generalists, organisms seen on both hard and soft substrata.
doi:10.1371/journal.pone.0044114.t001

Species Diversity and Community Composition at Background Sites

Univariate analysis showed no significant difference in diversity (density, S, H', J) between the R sites and Pre sites (assessing temporal variation between 2006 and 2009). There was also no significant difference in multivariate community composition among the background (Pre and R) transects (PERMANOVA, $F_{(1)} = 1.405$, p = 0.171). However, assessing fine-scale spatial heterogeneity, there was a positive relationship between the number of rocks in the background transects and species richness and diversity (linear regression; Rarefied species richness: $R^2 = 0.51$, ANOVA, $F_{(1, 15)} = 15.58$, p<0.001, Shannon Wiener species diversity: $R^2 = 0.60$, ANOVA, $F_{(1, 15)} = 22.47$, p<0.001; Figure 3). Rocks were unevenly distributed throughout the survey area and their presence increased the between-transect variation in measures of total density and diversity.

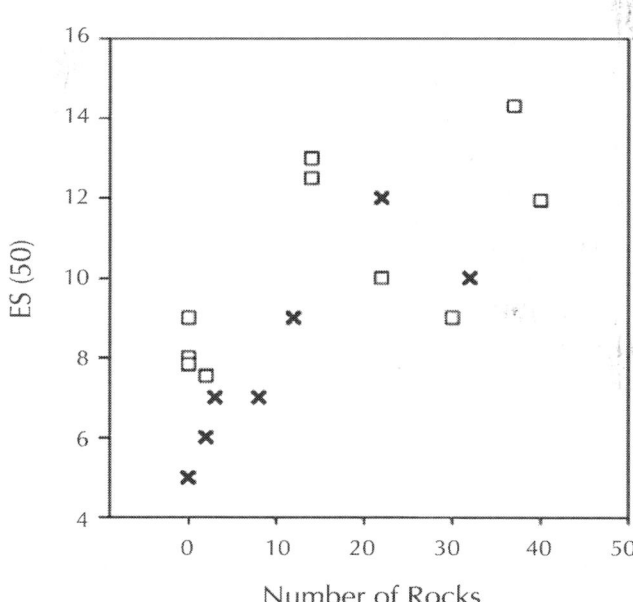

(a)

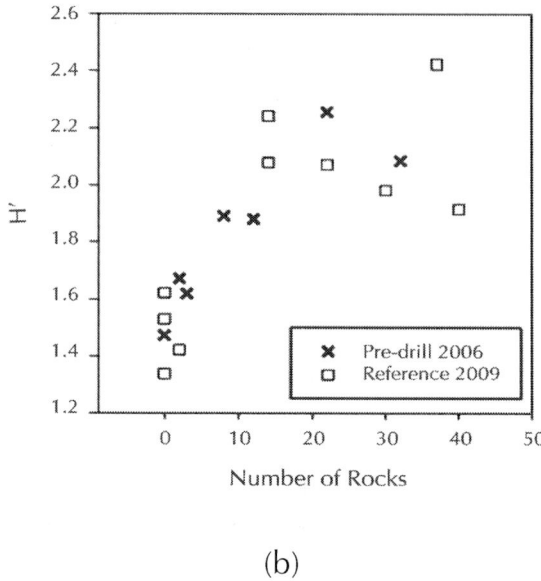

(b)

Figure 3: Habitat heterogeneity and species diversity at background sites. The relationship between the number of rocks observed in video transects and two indices of megafaunal invertebrate species diversity for the 2006 Pre-drill video survey and 2009 Reference sites (randomly selected undisturbed locations) (left; rarefied species richness $ES_{(50)}$, right; Shannon-Weiner Index H').

Physical Disturbance

The well was drilled in April 2006, resulting in the discharge of 192000 kg of barite drilling mud to the seabed. Discharge to the seabed was only from the top-hole (42″ and 36″ diameter sections). In addition, 77000 kg of barite were discharged to the sea surface from the 17.5″ section (Aas 2008, unpublished report). This resulted in a disturbance to the seabed with visible cuttings extending beyond 100 m in some directions in the two post-drilling surveys in 2006 (Figure 4). The visible extent of this disturbance decreased from >26600 m² in 2006 to 3500 m² by 2009. Seventy-six days after drilling, the cuttings reached 400 mm in thickness close to the well. At 50 m distance the thickness of the deposit was considerably less

(<50 mm) but still evident as a layer at the surface of the push cores (Table 2). In Post drill 2 the mean sediment barium (Ba) concentration (5450 mg kg⁻¹) was elevated above the pre-drilling concentration (150 mg kg⁻¹) and Norwegian Continental Shelf background levels (4.6–554 mg kg⁻¹) (SINTEF unpublished report). Three years later (Post 3), the mean sediment surface Ba concentration remained high at 25 and 50 m from the well (6133 and 6291 mg kg⁻¹ in the top 20 mm) but decreased with depth in the sediment (3283 and 547 mg kg⁻¹ at 40–60 mm). There were significant differences in the Ba concentration of the sections taken from different depths in the sediment at both 25 m (ANOVA, $F_{(2, 21)} = 4.02$, $p<0.05$) and 50 m (Kruskal-Wallis $H_{(2)} = 15.97$, $p<0.001$). Sediment particle size was affected by the deposition of drilling mud and cuttings. The percentage of particles under 69 μm (% fines) increased in the near-well samples taken in Post 2 in comparison to the Pre samples and those taken further from the disturbance in Post 2. In Post 3, the % fines remained high in the surface sediment at both 25 and 50 m from the well. There was a reduction in % fines with depth in the sediment, reaching values similar to Pre-drill at 20–40 and 40–60 mm depth at 50 m from the well.

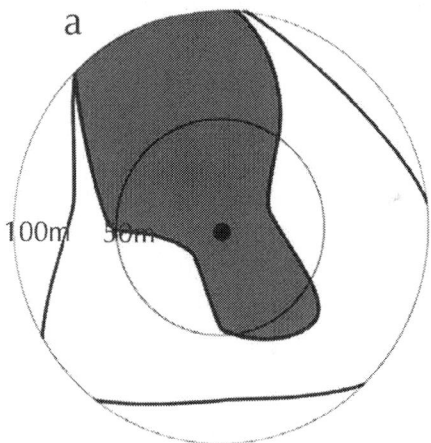

(a)

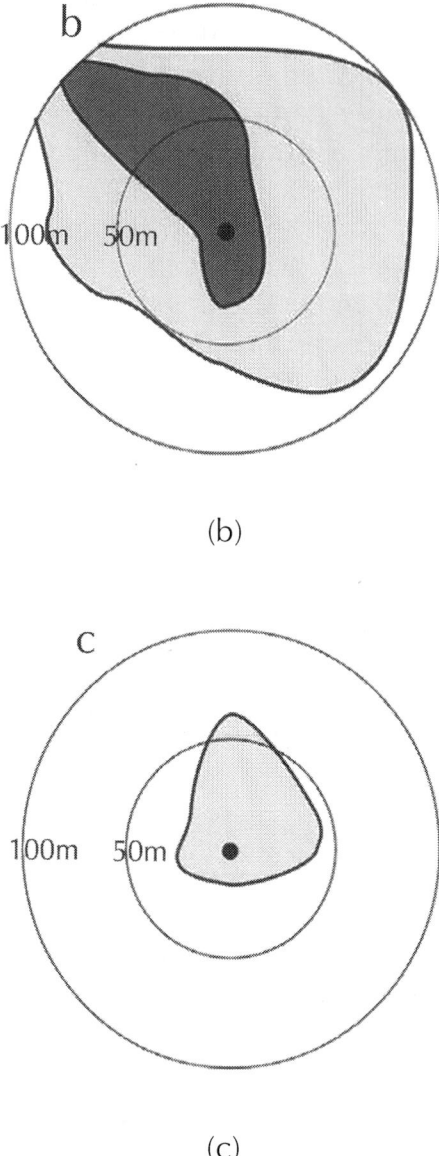

(b)

(c)

Figure 4: Physical disturbance at Morvin. Representation of the visible horizontal extent of drilling disturbance at Morvin: a) Post 1, b) Post 2, c) Post 3. The filled black circles in the centre represent well position, dark grey = complete coverage of sea bed with drill cuttings; light grey = partial coverage.

Table 2: Measurements of the depth of drill cuttings from graduated marker buoys and Barium concentration and sediment particle size from push core samples taken at the Morvin site during the Pre, Post 2 and Post 3 surveys

Survey	Section (mm)	Distance (m) from well	Depth of cuttings (mm)	Ba (mg kg2¹)	Sediment particle size; % fines (,69 mm
pre	0–50	0	0	150	53.6
pre	0–50	50	0	n/a	37.9
pre	0–50	100	0	n/a	38.4
Post2	0–50	0-10	400	5450 (61202)	80.2 (63.4)
Post2	0–50	100	<50	230	45.4 (64.3)
Post3	0–20	25		6133(61332)	76.9 (615.6)
Post3	20-40	25		4791(61998)	63.8 (626.9)
Post3	40-60	25		3283(62525)	60.3 (627.9)
Post3	0-20	50		6291(61505)	58.1 (619.7)
Post3	20-40	50		1991(62438)	41.3 (69.6)
Post3	40-60	50		547(6454)	37.1 (66.0)

Effects of Disturbance on Megafaunal Assemblage Composition

There was variation in mean density of soft-substrate, hard-substrate and generalist megafauna between the sampling units (Figure 5). There was a significant main effect of distance (L-ratio$_{(d.f. = 5)}$= 27.703, p<0.001) and time (L-ratio$_{(3)}$ = 25.362, p<0.001) on total density of benthic invertebrate megafauna at Morvin. There was, however, no significant interaction (L-ratio$_{(3)}$ = 0.634, p = 0.889). Soft sediment invertebrate megafaunal density showed an effect of distance (L-ratio$_{(5)}$ = 15.6, p<0.01) but no significant effect of time (L-ratio$_{(3)}$ = 6.195, p = 0.1). There was no significant interaction (L-ratio$_{(3)}$ = 0.785, p = 0.85). For the density of hard substrate invertebrates there was a significant main effect of time (L-ratio$_{(3)}$ = 13.467, p =

0.004), but no significant effect of distance (L-ratio$_{(5)}$ = 3.985, p = 0.552) or interaction between distance and time (L-ratio$_{(3)}$ = 4.03, p = 0.258).

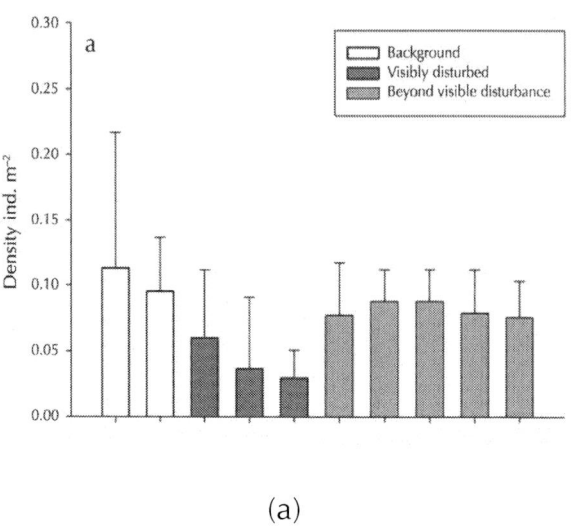

(a)

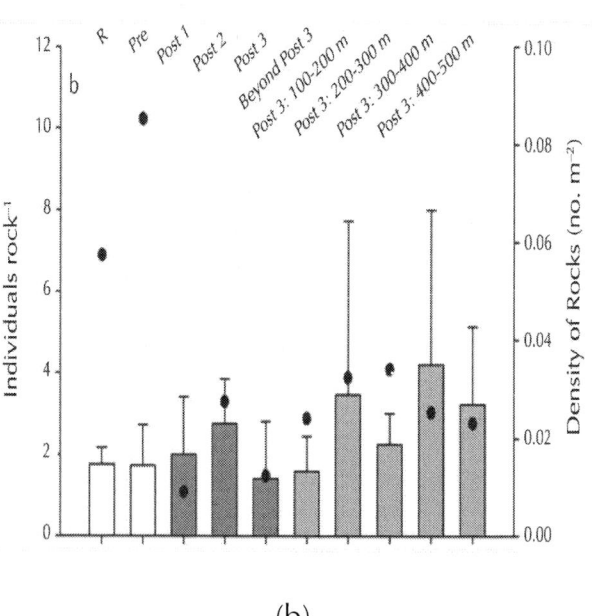

(b)

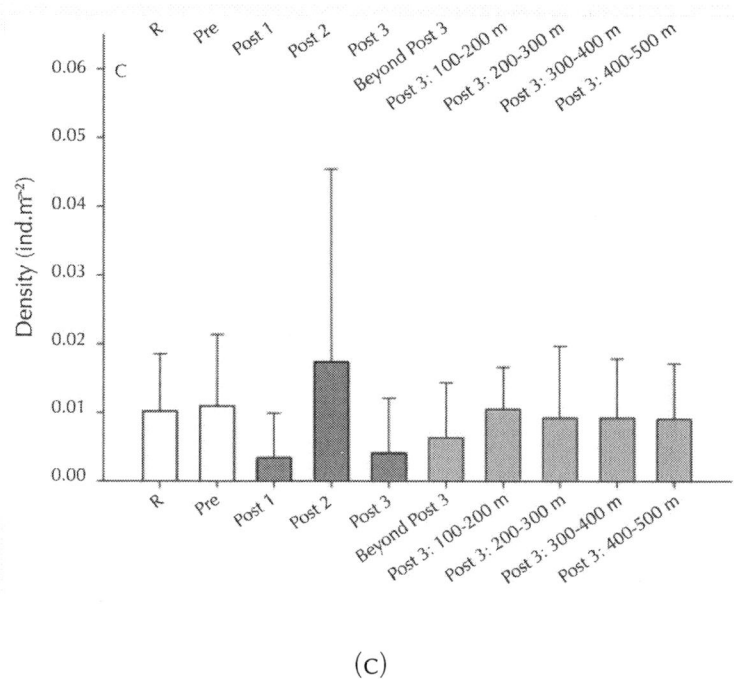

(c)

Figure 5: Mean (±sd) megafaunal invertebrate density (individuals m⁻²) at Morvin. (a) soft sediment, (b) hard substrate and (c) generalist megafauna. Background sites are shown in white, visibly disturbed areas in dark grey and areas beyond disturbance are shown in light grey. Filled circles in hard substratum chart present show the density of rocks in the transects.

PERMANOVA showed a significant effect of time ($F_{(3)} = 0.163$, $p < 0.001$) but no significant effect of distance ($F_{(5)} = 0.055$, $p = 0.163$) or interaction ($F_{(3)} = 0.021$, $p = 0.761$). The multidimensional scaling plot (Figure 6) of the combined data for each transect disturbance/distance zone grouped the R and Pre sites and the sites beyond disturbance at the 80% similarity level.

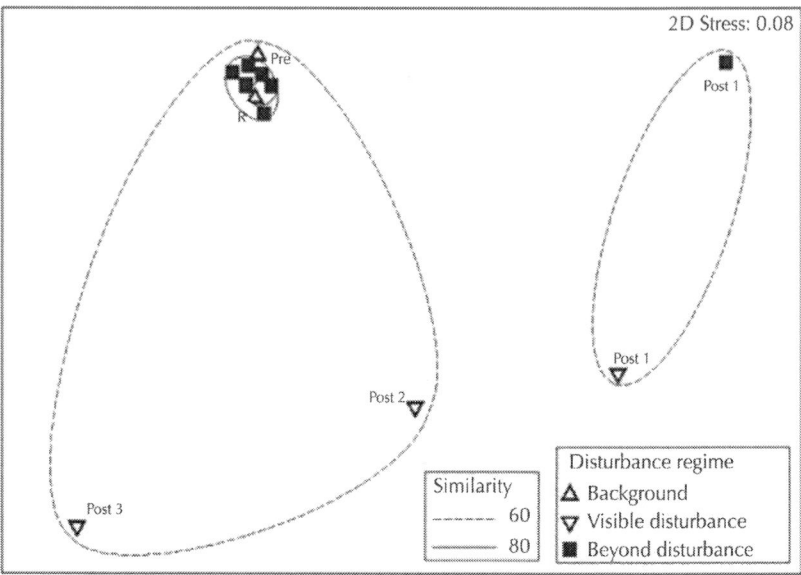

Figure 6: Multidimensional scaling ordination of megafaunal assemblages under different disturbance conditions. Based on Bray Curtis similarity of pooled invertebrate megafaunal density data for the disturbance zones in 2006 (Pre, Post 1 and Post 2) and 2009 (Post 3, R). For each survey the transects have been divided into Background, Visible Disturbance and Beyond Disturbance according to the coverage of the sediment by drill cuttings, notable groups are labelled. Similarity levels from cluster analysis.

It should be noted that the two-way design used here was limited by the lack of samples from distance zones greater than 100 m in all years except Post 3. There were only two distance zones for samples at most times, both within 100 m from the drilling activity. This limited replcation will reduce the ability to detect a main effect of distance or an interaction between distance and time in the statistical tests.

Recovery

For each of the indices tested, the transformed percentage difference from Pre-drill (Response Y) varied across the distance and time scales considered (Figure 7), but was generally more negative close to the disturbance event in both space and time. Response Y for the density of motile organisms showed no main effects of distance (ANOVA on ranks $F_{(5,70)} = 2.135$, $p = 0.071$), time ($F_{(2,70)} = 1.253$, $p = 0.292$) or the interaction ($F_{(2,70)} = 1.297$, $p = 0.280$). Response Y for the density of sessile organisms showed significant main effects of distance ($F_{(5,70)} = 3.967$, $p<0.01$), but no significant effect of time ($F_{(2,70)} = 1.928$, $p = 0.153$) or the interaction ($F_{(2,70)} = 0.702$, $p = 0.499$). Response Y for the Shannon-Wiener diversity and estimated richness (ES_{50}) of megafauna revealled significant main effects of distance (H': $F_{(5,70)} = 14.116$; ES_{50}: $F_{(5,70)} = 16.530$; $p<0.001$ for both) and time (H': $F_{(2,70)} = 4.947$, $p<0.01$; ES_{50}: $F_{(2,70)} = 4.027$, $p<0.05$) and the interaction (H': $F_{(2,70)} = 3.349$, $p<0.05$; ES_{50}: $F_{(2,70)} = 3.280$, $p<0.05$). Response Y for the evenness of megafauna (J) had a significant main effect of distance ($F_{(5,70)} = 4.275$, $p<0.05$) but no significant main effect of time ($F_{(2,70)} = 0.640$, $p = 0.530$) or the interaction ($F_{(2,70)} = 0.700$, $p = 0.502$). Response Y for the Bray-Curtis similarity between megafaunal assemblages did not reveal any significant effects of distance ($F_{(5,70)} = 2.068$, $p = 0.080$), time ($F_{(2,70)} = 2.551$, $p = 0.085$) or the interaction ($F_{(2,70)} = 0.310$, $p = 0.734$).

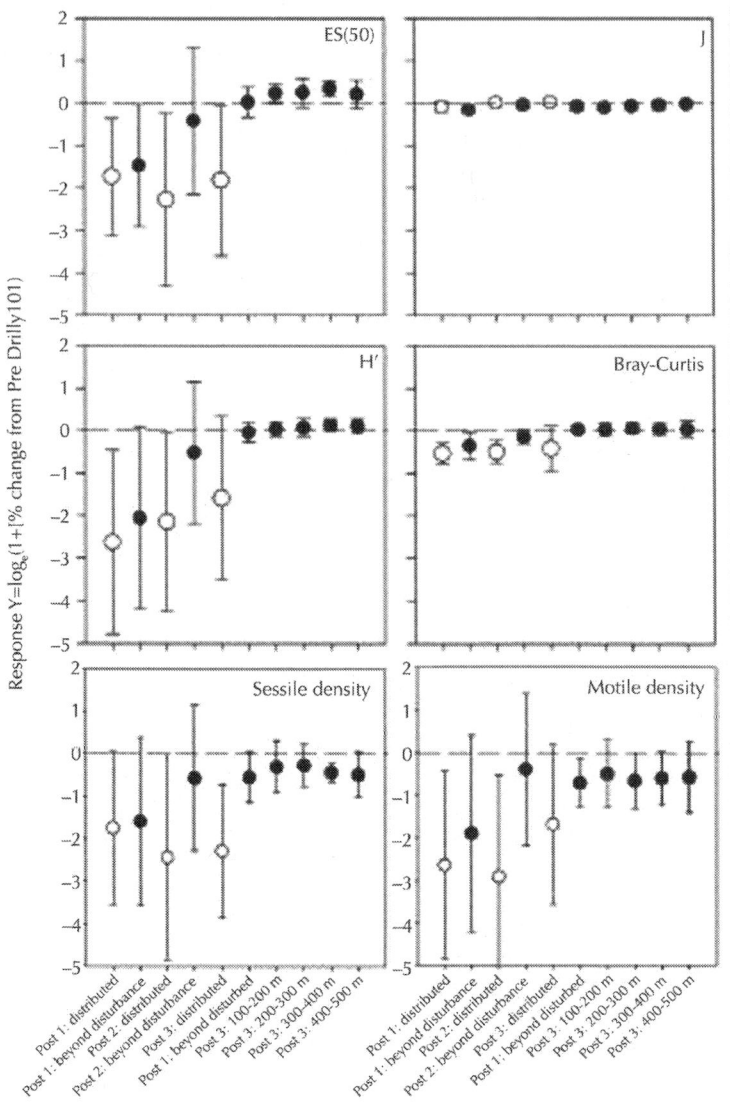

Figure 7: Response Y recovery index in comparison to pre-drill. Shown for Rarefied species richness ($ES_{(50)}$), Species evenness (*J*), Shannon-Wiener Index (*H'*), Bray Curtis similarity, total sessile organism density and total motile organism density. Unfilled circles indicate disturbed zones and filled circles indicate distance from disturbance. Dashed lines indicate zero. Error bars = standard deviation.

Evidence of Biological Activity

Decapod burrows were common in the soft sediment at Morvin with mean densities of 3.5 m^{-2} in the Reference sites. Mean decapod burrow density differed significantly along the disturbance gradient in the 2009 surveys (Figure 8; ANOVA $F_{(6, 51)}$ = 4.77, p<0.001). Pairwise comparisons (Holm-Sidak method) showed significant differences between the 2009 disturbed zone and all the other zones with the exception of the undisturbed area within 100 m of the well. The closest burrow was 5 m from the well and the numbers began to increase after 20 m distance.

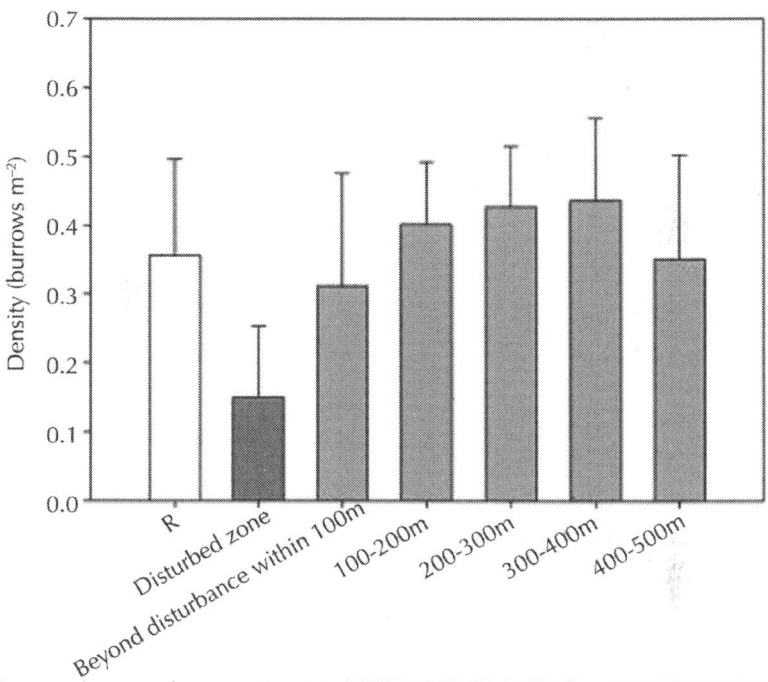

Figure 8: Mean density of decapod burrows (±sd) in the Post 3 survey at Morvin. White = Background, dark grey = Visible Disturbance, light grey = Beyond Disturbance.

DISCUSSION

Background Environment

At Morvin the rocks on the seabed provided heterogeneity in an otherwise soft-sediment environment. This is a typical situation for the northern North Atlantic [45]. The increase in habitat heterogeneity enhanced benthic diversity at Morvin, as has been shown elsewhere at global [46] and local scales [47]. The composition of the soft-sediment megafaunal assemblage was comparable to that found in areas of similar depth in the Porcupine Seabight (Table 3), southwest of Ireland, with species densities also similar [48], [49], [50], [51]. The available hard substratum at Morvin increased habitat heterogeneity with a resultant increase in density and species richness, most notably in the Porifera. In this respect there are direct similarities between Morvin and the megafauna from the Faroe-Shetland Channel to the south [52] and Le Danois Bank at equivalent depth in the Cantabrian Sea (Table 3) where exposed rock on an otherwise sandy seabed formed a distinct habitat with high abundances of the sponge *Phakellia ventilabrum* [53]. Of the demersal fish at Morvin, the presence of *Lophius piscatorius*, *Sebastes* sp. and *Chimera monstrosa* were consistent with results from previous studies of Norwegian shelf-edge Atlantic water [54].

Table 3: Mean taxon densities (numbers 100 m^{-2}) of species shared between Morvin and other Atlantic sites at similar depth

Taxon	Morvin	Le Danois Bank	Porcupine Seabight	West of Shetland
Hymedesmia sp	0.42	5.67		0.73
Phakellia sp.	0.08	1.02		
Kophobelemnon stel-liferum	5.63		260260	
Funiculina sp.	1.74	0.04(T)		

Cerianthus sp.	0.39	0.638		
Colus sp.	0.14	0.22(T)	X	
Geryon sp.	0.28	0.03(T)	X	0.72
Pandalus sp.	0.42		X	1.65
Munida sp.	0.14	0.829	X	7.91
Parastichopus tremulus	2.62	0191	X	7.69
Henricia sp.	0.2		X	1.61
Ceramaster sp.	0.26			1.32
Echinus sp.	0.14	0.04(T)		2.93
Asterias rubens	X		x	0.29
Porania sp.	0.2	0.32		0.55
Cidaris cidaris	0.14	0.01		11.73
No of taxa in common with Morvin		11	8	11

It is important to evaluate disturbance-related changes at Morvin within the context of the broader temporally-dynamic ecosystem. Temporal change in the deep sea is receiving increasing attention [55] and recent studies have shown seasonal and interannual changes in benthic megafaunal communities in the deep north-east Atlantic [29], [56], [57], [58]. Shallow-water studies of recovery trajectories have identified seasonal recruitment as an important factor [59]. At Morvin the megafauna in the Pre-drill (23rd March 2006) and R site surveys (3–4th May 2009) showed no significant differences in density, diversity or assemblage composition. This suggests that there was limited natural temporal change at the time-scale analysed, thus supporting comparisons between pre- and post-drilling

Initial Disturbance

Recent studies in the north-east Atlantic have revealed drill cuttings extending to approximately 200 m from the well with reduced megafaunal density and diversity within the disturbed area[4], [15]. In the present case, the visible extent of the cuttings reached beyond 100 m from the BOP to the north-west but were generally less than 100 m. This equates to an area of at least 26601 m². This is considerably smaller than reported in older studies of exploration wells in the north-east Atlantic in which oil based drilling mud was used and there was less regulation for discharges to the seabed [20], [60]. The extent was also smaller than reported in more recent studies at a similar depth in the Faroe-Shetland Channel (>66800 m²) [4], albeit at a site with multiple wells drilled. The persistance of the effects of water based mud and drill cuttings on megafauna is unknown and the increasing number of wells in a field could result in larger areas being affected, with potential accumulating or synergistic long-term effects.

The drill cuttings deposited at Morvin caused an initial physical disturbance, which resulted in smothering of the benthic fauna. The longer-term impacts associated with such an event include the possible chemical effects of the drilling mud, hypoxia related to chemicals in the drilling mud or to smothering, and reduced habitat heterogeneity caused by the rapid creation of a smooth, soft-sediment environment. In terms of the physical nature of the disturbance caused by exploration drilling, there are similarities with the disposal of dredged material [61] and bottom trawling [62], [63].

Persistence of the Disturbance

Although there was still visible evidence of disturbance surrounding the well in 2009, the total area visibly disturbed by cuttings deposition had decreased considerably since 2006. Then, the cuttings pile was over 400 mm deep at 10 m distance from the well and at 50 m there was a thin covering of unevenly distributed drill cuttings, estimated to be less than 50 mm. Although the area of deeper cuttings coverage was the most impacted area in this study, the area with the thinner layer of cuttings can not be discounted as even a thin layer of cuttings may affect the sediment bacteria and smaller size fractions of benthic fauna. These organisms were not visible in the video methods used in this study but have important roles in the functioning of benthic ecosystems as well as providing food source to some megafaunal organisms. An elevated "crater" remained at the exact well location which attracted increased abundance of the fish *Sebastes* sp. (excluded from the quantitative analysis). The increased quantity of cuttings deposited close to the well, and the cement used to secure the structure of the well in the plug and abandon phase [64], may consolidate the cuttings pile in the immediate vicinity of the well. It has been suggested that, unless disturbed, cuttings piles remain relatively unchanged over time [10] and that the cuttings further from the well may be stable [65]. As a result, the obscurring of the disturbed sediment by the natural settlement of material from the water column may be a more important factor in reducing the visible extent of the cuttings than the erosion and lateral transport of the deposited drill cuttings by the currents. Indeed, large accumulations of sediment on coral reefs in the Morvin area [34] suggest relatively high sedimentation rates. However, lateral transport and the resulting breakdown of cuttings piles has been suggested by the presence of barite particles incorporated into the skeletons of corals located 4 km away from a 20 year old exploration well elsewhere in the Norwegian Sea [66].

Barium levels at Morvin were elevated, indicating persistance of the drill cuttings after three years. Although Ba is considered non-toxic, there remains debate in relation to the use of barite as a weighting agent in drilling mud. A variety of sublethal effects have been reported from laboratory studies such as reduced condition (gill damage) in benthic bivalves [67] and lower colonization by macrofauna of sediment treated with barite [68]. Other studies suggest the deposition of barite results in changed physical properties of the sediment [8], which in turn may alter habitat heterogeneity and increase meiofaunal density, as shown in a laboratory study[69]. The most abundant motile organism at Morvin, the holothurian *Parastichopus tremulus*, was completely absent from the disturbed areas of the post-drilling surveys in 2006. Seasonal variations in the density of *P. tremulus* are known [70] but owing to the relatively short time period between the Pre-drill survey and the first Post-drill survey, and consistent abundance of *P. tremulus* at the same time of year in 2009 outside the disturbed zone, it is likely that this species was absent because of the disturbance. This could be either because holothurian distribution is determined by food particle availability [51], which may be reduced on the newly deposited cuttings, or because holothurians ingest food particles selectively [71] and may therefore avoid consuming the cuttings which consist of differing physical properties [8] to the background sediment.

Megafaunal Recovery

There does not appear to be differential recovery between the visible disturbance zones within 100 m of the well (an interaction between distance and time factors), although these tests were limited by low replication. However, at a finer scale abundance was still reduced in the immediate vicinity of the well in the Post 3 survey.

Within 100 m of the drilling there were detectable differences in total megafauna between the visibly disturbed and not visibly disturbed areas. Most of this variation appeared to be within the sessile fauna. In comparison to the sites further from disturbance in 2009 there was increased variability in the samples close to the source of disturbance both spatially and temporally. Increased variability has been discussed as an indication of stress in marine communities [72]. In terms of the benthic megafauna, the most notable difference in the community structure between the 2009 disturbed zone and the reference sites was the reduction in sessile organisms. After the drilling operations, the dominant sponges on the hard substrata (*Phakellia*sp. and *Mycale* sp.) were rare, primarily because of burial of their habitat. Further research is required to determine how sponges respond to lower degrees of sedimentation leading to partial burial. Throughout the study pennatulids were the most common organisms on the soft sediment. Their numbers were low in the visibly disturbed area in 2009. Pennatulids are slow growing and may therefore take considerable time to recover from disturbance [73]. The larval recruitment and settlement rates for these organisms are unknown. Studies on the reproduction of *Kophobelemnon stelliferum*, *Pennatula phosphorea* and *Funiculina quadrangularis* suggest these species have lecithotrophic larvae, which may remain in the water column until suitable habitat is located [49], [74], [75] and could possibly avoid settlement on sediment disturbed by drilling mud and cuttings.

Bioturbation rates are poorly understood in deep water but are important indicators of ecosystem function. This process is evidently important in the recovery of soft sediments after physical disturbance. In the Post 3 survey, large burrows were present on the disturbed seabed, indicating activity of the decapod *Geryon* sp. in this area. These crabs were observed entering and leaving these burrows, the structure of which was very similar to *Geryon trispinosus* burrows on the seafloor of the Porcupine Seabight [48]. This activity is likely to be important in the re-distribution of the sediment and gradual breakdown of the cuttings pile. The nearest burrow was 5 m from the well indicating activity in this area in the three years since disturbance. The presence of new burrows and the apparent longevity of some *Lebenspuren*[76] implies that reduced burrow density may not necessarily indicate long-term reduction in bioturbation activity. The holothurian *Parastichopus tremulus* is important in horizontal dispersal of sediment [51] and therefore, potentially, in the re-distribution of cuttings and drilling mud. However, the Morvin data suggest *P. tremulus* avoids feeding on the cuttings and thus probably does not contribute much to the re-distribution of sediments. Although not considered in this study, the inclusion of the macrofauna, which may be more abundant than the megafauna both numerically and in terms of biomass and which include important bioturbators, would benefit future studies of recovery. Indeed, experimental data suggest that macrofaunal assembalges may colonize water based drilling mud rapidly [77]. In addition, the chemical disturbance and altered sediment characteristics may also affect meiofaunal assemblage composition [78], [79]and the microbial assemblage, which could influence food availability and therefore the recovery of the larger benthic fauna.

Studies on the Georges Bank, Gulf of Maine (60–100 m depth) suggest limited effects of oil and gas exploration activities on megafauna (at finer-scale resolution than Morvin) and evidence of recovery by the macrofauna [80]. The Georges Bank is subject to high energy storms that redistribute sediments. In contrast, at a lower energy abyssal site experimental disturbances designed to predict the effects of nodule mining [23] showed limited evidence for recovery of the megafauna after seven years with no subsequent disturbance. It has been suggested that recovery is complicated and influenced by factors including the scale of the disturbance [81], the type and frequency of disturbance and the local environmental conditions [62], [82]. These factors complicate the assessment of recovery in studies such as this one, limited by operational contraints (access to a deep site, spatial reach of the ROV in the earlier surveys) and highlight the importance of suitable spatial and temporal replication. To address this issue, bioequivalence methods have been used to assess ecological impacts [83] but have not been universally adopted in ecological studies [84].

The limited and ambiguous data on benthic recovery in deeper water highlight the need for more studies. At present, differences in the physical and biological environments at different study sites and the individual nature of each cuttings pile make it impossible to draw general conclusions. A similar study of a drilling site in the Faroe-Shetland Channel [85] has also revealed a small area of reduced faunal density and diversity close to the well after three years. We suggest that the significant decrease in megafaunal density, which appears to persist for at least 3 years at both sites will occur at all deep-water drilling sites, with the severity of the impact likely to be correlated with the amount of material deposited on the seabed and the local environmental conditions. It is anticipated that the effect will be greater in deeper, colder areas, where the rate of metabolism and growth are expected to be considerably lower [76], thereby reducing the rate of recovery. The change in sediment particle size may also retard recovery, as demonstrated in shallower water [86]. With increasing anthropogenic activity in deeper waters it is essential to understand the initial effects on benthic fauna and their recovery to such impacts. Hydrocarbon exploration disturbance provides a valuable tool to study disturbance and recovery trajectories in remote deep-water habitats, which are generally difficult to access.

REFERENCES

1. Gage JD (2001) Deep-sea benthic community and environmental impact assessment at the Atlantic Frontier. Continental Shelf Research 21: 957–986.

2. Pinder D (2001) Offshore oil and gas: global resource knowledge andtechnological change. Ocean & Coastal Management 44: 579–600.

3. Kotchen MJ, Burger NE (2007) Should we drill in the Arctic National Wildlife Refuge? An economic perspective. Energy Policy 35: 4720–4729

4. Jones DOB, Hudson IR, Bett BJ (2006) Effects of physical disturbance on the cold-water megafaunal communities of the Faroe-Shetland Channel. Marine Ecology Progress Series 319: 43–54.

5. Hylland K, Tollefsen KE, Ruus A, Jonsson G, Sundt RC, et al. (2008) Water column monitoring near oil installations in the North Sea 2001–2004. Marine Pollution Bulletin 56: 414–429.

6. Crone TJ, Tolstoy M (2010) Magnitude of the 2010 Gulf of Mexico Oil Leak. Science 330: 634–634.

7. Holdway DA (2002) The acute and chronic effects of wastes associated with offshore oil and gas production on temperate and tropical marine ecological processes. Marine Pollution Bulletin 44: 185–203.

8. Schaanning MT, Trannum HC, Øxnevad S, Carroll J, Bakke T (2008) Effects of drill cuttings on biogeochemical fluxes and macrobenthos of marine sediments. Journal of Experimental Marine Biology and Ecology 361: 49–57.

9. Smit MGD, Holthaus KIE, Trannum HC, Neff JM, Kjeilen-Eilertsen G, et al. (2008) Species sensitivity distributions for suspended clays, sediment burial, and grain size change in the marine environment. Environmental Toxicology and Chemistry 27: 1006–1012.

10. Breuer E, Stevenson AG, Howe JA, Carroll J, Shimmield GB (2004) Drill cutting accumulations in the Northern and Central North Sea: a review of environmental interactions and chemical fate. Marine Pollution Bulletin 48: 12–25.

11. Hartley JP (1996) Environmental monitoring of offshore oil and gas drilling discharges-A caution on the use of barium as a tracer. Marine Pollution Bulletin 32: 727–733.

12. Netto SA, Gallucci F, Fonseca G (2009) Deep-sea meiofauna response to synthetic-based drilling mud discharge off SE Brazil. Deep-Sea Research Part IITopical Studies in Oceanography 56: 41–49.

13. Currie DR, Isaacs LR (2005) Impact of exploratory offshore drilling on benthic communities in the Minerva gas field, Port Campbell, Australia. Marine Environmental Research 59: 217–233.

14. Santos MFL, Lana PC, Silva J, Fachel JG, Pulgati FH (2009) Effects of nonaqueous fluids cuttings discharge from exploratory drilling activities on the deepsea macrobenthic communities. Deep-Sea Research Part II-Topical Studies in Oceanography 56: 32–40.

15. Jones DOB, Wigham BD, Hudson IR, Bett BJ (2007) Anthropogenic disturbance of deep-sea megabenthic assemblages: a study with RemotelyOperated Vehicles in the Faroe-Shetland Chanel, NE Atlantic. Marine Biology 151: 1731–1741.

16. Nodder SD, Pilditch CA, Probert PK, Hall JA (2003) Variability in benthic biomass and activity beneath the Subtropical Front, Chatham Rise, SW Pacific Ocean. Deep Sea Research Part I: Oceanographic Research Papers 50: 959– 985.

17. Danovaro R, Gambi C, Dell'Anno A, Corinaldes C, Fraschetti S, et al. (2008) Exponential decline of deep-sea ecosystem functioning linked to benthic diversity loss Current Biology 18: 1–8.

18. Trannum HC, Nilsson HC, Schaanning MT, Norling K (2011) Biological and biogeochemical effects of organic matter and

drilling discharges in two sediment communities. Marine Ecology Progress Series 442: 23–36.

19. Trannum HC, Nilsson HC, Schaanning MT, Oxnevad S (2010) Effects of sedimentation from water-based drill cuttings and natural sediment on benthic macrofaunal community structure and ecosystem processes. Journal of Experimental Marine Biology and Ecology 383: 111–121.

20. Olsgard F, Gray JS (1995) A Comprehensive Analysis of the Effects of Offshore Oil and Gas Exploration and Production on the Benthic Communities of the Norwegian Continental-Shelf. Marine Ecology Progress Series 122: 277–306.

21. Davies JM, Bedborough DR, Blackman RAA, Addy JM, Appelbee JF, et al. (1989) Environmental effects of oil-based mud drilling in the North Sea. In: Englehardt FR, Ray JP, Gillam AH, editors. Drilling wastes. London: Elsevier Applied Science.

22. Mair JMD, Matheson I, Appelbee JF (1987) Offshore macrobenthic recovery in the Murchison field following termination of drill cuttings discharge. Marine Pollution Bulletin 18: 628–634.

23. Bluhm H (2001) Re-establishment of an abyssal megabenthic community after experimental physical disturbance of the seafloor. Deep-Sea Research Part II: Topical Studies in Oceanography 48: 3841–3868.

24. Bluhm H, Schriever G, Thiel H (1995) Megabenthic recolonization in an experimentally disturbed abyssal manganese nodule area. Marine Georesources & Geotechnology 13: 393–416.

25. O'Neill RV (1998) Recovery in complex ecosystems. Journal of Aquatic Ecosystem Stress and Recovery 6: 181–187.

26. Grassle JP, Sanders RR, Hessler GT, Rowe GT, McLellan T (1975) Pattern and zonation: a study of the bathyal megafauna using the research submersible Alvin. Deep-Sea Research 22: 457–481.

27. Smith CR, Hamilton SC (1983) Epibenthic megafauna of a bathyal basin off southern California: patterns of abundance, biomass, and dispersion. Deep-Sea Research 30: 907–928.

28. Vardaro MF, Ruhl HA, Smith KL (2009) Climate variation, carbon flux, and bioturbation in the abyssal North Pacific. Limnology and Oceanography 54: 2081–2088.

29. Bett BJ, Malzone MG, Narayanaswamy BE, Wigham BD (2001) Temporal variability in phytodetritus and megabenthic activity at the seabed in the deep Northeast Atlantic. Progress in Oceanography 50: 349–368.

30. Beaulieu SE (2001) Life on glass houses: sponge stalk communities in the deep sea. Marine Biology 138: 803–817.

31. Hughes SJM, Jones DOB, Hauton C, Gates AR, Hawkins LE (2010) An assessment of drilling disturbance on Echinus acutus var. norvegicus based on in-situ observations and experiments using a remotely operated vehicle (ROV). Journal of Experimental Marine Biology and Ecology 395: 37–47.

32. Larsson AI, Purser A (2011) Sedimentation on the cold-water coral Lophelia pertusa: Cleaning efficiency from natural sediments and drill cuttings. Marine Pollution Bulletin 62: 1159–1168.

33. Jones DOB (2009) Using existing industrial remotely operated vehicles for deepsea science. Zoologica Scripta 38: 41–47.

34. Hovland M (2008) Deep-water coral reefs: Unique biodiversity hotspots. New York: Springer-Praxis. 278 p.

35. Hovland M, Heggland R, De Vries MH, Tjelta TI (2010) Unit-pockmarks and their potential significance for predicting fluid flow. Marine and Petroleum Geology 27: 1190–1199. Recovery of Megafauna from Drilling Disturbance

36. Magurran AE (2003) Measuring Biological Diversity. Oxford: Blackwell Science. 260 p.

37. Clarke KR, Warwick RM (2001) Changes in Marine Communities: An Approach to Statistical Analysis and Interpretation: Plymouth Marine Laboratory, U.K. 205 p.

38. Dobson AJ, Barnett AG (2008) An Introduction to Generalized Linear Models. 3rd Edition. London: Chapman & Hall.

39. Fox J, Weisberg S (2011) An R Companion to Applied Regression, Second Edition. Thousand Oaks CA: Sage.

40. R Development Core Team (2010) R: A language and environment for statistical computing. R Foundation for Statistical Computing. ISBN 3-900051-07-0, URL http://www.R-project.org., Vienna, Austria.

41. Bray JR, Curtis JT (1957) An ordination of the upland forest of southern Wisconsin. Ecological Monographs 27: 225–349.

42. Anderson MJ (2001) A new method for non-parametric multivariate analysis of variance. Austral Ecology 26: 32–46.

43. Oksanen J, Blanchet FG, Kindt R, Legendre P, O'Hara RB, et al. (2011) Vegan: Community Ecology Package. R package version 1.17-9. http://CRAN.Rproject.org/package = vegan.

44. Kaiser MJ, Clarke KR, Hinz H, Austen MCV, Somerfield PJ, et al. (2006) Global analysis of response and recovery of benthic biota to fishing. Marine Ecology Progress Series 311: 1–14.

45. Schulz M, Bergmann M, von Juterzenka K, Soltwedel T (2010) Colonisation of hard substrata along a channel system in the deep Greenland Sea. Polar Biology 33: 1359–1369.

46. Vanreusel A, Fonseca G, Danovaro R, da Silva MC, Esteves AM, et al. (2010) The contribution of deep-sea macrohabitat heterogeneity to global nematode diversity. Marine Ecology: an Evolutionary Perspective 31: 6–20.

47. Buhl-Mortensen L, Vanreusel A, Gooday AJ, Levin LA, Priede IG, et al. (2010) Biological structures as a source of habitat heterogeneity and biodiversity on the deep ocean margins. Marine Ecology: an Evolutionary Perspective 31: 21–50.

48. MJ, Hartnoll RG, Rice AL (1991) Aspects of the biology of the deep-sea crab Geryon trispinosus from the Porcupine Seabight. Journal of the Marine Biological Association of the United Kingdom Plymouth 71: 311–328.

49. Rice AL, Tyler PA, Paterson GJL (1992) The pennatulid Kophobelemnon stelliferum (Cnidaria, Octocorallia) in the

Porcupine Seabight (Noth-East Atlantic Ocean). Journal of the Marine Biological Association of the United Kingdom 72: 417– 434.

50. Rice AL, Aldred RG, Darlington E, Wild RA (1982) The quantitative estimation of the deep-sea megabenthos - a new approach to an old problem. Oceanologica Acta 5: 63–72.

51. Billett DSM (1991) Deep-sea holothurians. Oceanography and Marine Biology: An Annual Review 29: 259–317.

52. Jones DOB, Bett BJ, Tyler PA (2007) Megabenthic ecology of the FaroeShetland Channel: a photographic study. Deep Sea Research Part I: Oceanographic Research Papers 54: 1111–1128.

53. Sa´nchez F, Serrano A, Go´mez Ballesteros M (2009) Photogrammetric quantitative study of habitat and benthic communities of deep Cantabrian Sea hard grounds. Continental Shelf Research 29: 1174–1188.

54. Bergstad OA, Bjelland O, Gordon JDM (1999) Fish communities on the slope of the eastern Norwegian Sea. Sarsia 84: 67–78.

55. Glover AG, Gooday AJ, Bailey DM, Billett DSM, Chevaldonne P, et al. (2010) Temporal change in deep-sea benthic ecosystems a review of the evidence from recent time-series studies. Advances in Marine Biology 58: 1–95.

56. Bergmann M, Soltwedel T, Klages M (2011) The interannual variability of megafaunal assemblages in the Arctic deep sea: Preliminary results from the HAUSGARTEN observatory (79uN). Deep Sea Research Part I: Oceanographic Research Papers 58: 711–723.

57. Billett DSM, Bett BJ, Reid WDK, Boorman B, Priede IG (2010) Long-term change in the abyssal NE Atlantic: The 'Amperima Event' revisited. Deep-Sea Research Part II: Topical Studies in Oceanography 57: 1406–1417.

58. Billett DSM, Bett BJ, Rice AL, Thurston MH, Galeron J, et al. (2001) Long-term change in the megabenthos of the Porcupine Abyssal Plain (NE Atlantic).

59. Progress in Oceanography 50: 325–348.

60. Kro¨ger K, Gardner JPA, Rowden AA, Wear RG (2006) Recovery of a subtidal soft-sediment macroinvertebrate assemblage following experimentally induced effects of a harmful algal bloom. Marine Ecology Progress Series 326: 85–98.

61. Davies JM, Hardy R, McIntyre AD (1981) Environmental effects of North Sea oil operations. Marine Pollution Bulletin 12: 412–416.

62. Foden J, Rogers SI, Jones AP (2009) Recovery rates of UK seabed habitats after cessation of aggregate extraction. Marine Ecology Progress Series 390: 15–26.

63. Engel J, Kvitek R (1998) Effects of otter trawling on a benthic community in Monterey Bay national marine sanctuary. Conservation Biology 12: 1204–1214.

64. Collie JS, Escanero GA, Valentine PC (1997) Effects of bottom fishing on the benthic megafauna of Georges Bank. Marine Ecology Progress Series 155: 159– 172.

65. Hyne NJ (2001) Nontechnical Guide to Petroleum Geology, Exploration, Drilling and Production, second edition. Tulsa: PennWell. 598 p.

66. Black KS, Paterson DM, Davidson IR (2002) Erosion of cuttings pile sediments: A laboratory flume study. Underwater Technology 25: 51–59.

67. Lepland A, Mortensen PB (2008) Barite and barium in sediments and coral skeletons around the hydrocarbon exploration drilling site in the Traena Deep, Norwegian Sea. Environmental Geology 56: 119–129.

68. Barlow MJ, Kingston PF (2001) Observations on the effects of barite on the gill tissues of the suspension feeder Cerastoderma edule (Linne) and the deposit feeder Macoma balthica (Linne). Marine Pollution Bulletin 42: 71–76.

69. Tagatz ME, Tobia M (1978) Effect of barite (BaSO4) on development of estuarine communities. Estuarine and Coastal Marine Science 7: 401–407.

70. Cantelmo FR, Tagatz ME, Rao KR (1979) Effect of barite on meiofauna in a flow-through experimental system. Marine Environmental Research 2: 301–309.

71. Sa´nchez F, Serrano A, Parra S, Ballesteros M, Cartes JE (2008) Habitat characteristics as determinant of the structure and spatial distribution of epibenthic and demersal communities of Le Danois Bank (Cantabrian Sea, N. Spain). Journal of Marine Systems 72: 64–86.

72. Hudson IR, Wigham BD, Tyler PA (2004) The feeding behaviour of a deep-sea holothurian, Stichopus tremulus (Gunnerus) based on in situ observations and experiments using a Remotely Operated Vehicle. Journal of Experimental Marine Biology and Ecology 301: 75–91.

73. Warwick RM, Clarke KR (1993) Increased variability as a symptom of stress in marine communities. Journal of Experimental Marine Biology and Ecology 172:215–226.

74. Wilson MT, Andrews AH, Brown AL, Cordes EE (2002) Axial rod growth and age estimation of the sea pen, Halipteris willemoesi Kolliker. Hydrobiologia 471: 133–142.

75. Edwards DCB, Moore CG (2008) Reproduction in the sea pen Pennatula phosphorea (Anthozoa : Pennatulacea) from the west coast of Scotland. Marine Biology 155: 303–314.

76. . Edwards DCB, Moore CG (2009) Reproduction in the sea pen Funiculina quadrangularis (Anthozoa: Pennatulacea) from the west coast of Scotland. Estuarine Coastal and Shelf Science 82: 161–168.

77. Gage JD, Tyler PA (1991) Deep-Sea Biology: A natural history of organisms at the deep-sea floor. Cambridge: Cambridge University Press. 504p.

78. Trannum HC, Setvik A, Norling K, Nilsson HC (2011) Rapid macrofaunal colonization of water-based drill cuttings on different sediments. Marine Pollution Bulletin 62: 2145–2156.

79. . Gray JS (2002) Species richness of marine soft sediments. Marine Ecology Progress Series 244: 285–297.

80. Schratzberger M, Warwick RM (1999) Differential effects of various types of disturbances on the structure of nematode assemblages: an experimental approach. Marine Ecology Progress Series 181: 227–236.

81. Neff JM, Bothner MH, Maciolek NJ, Grassle JF (1989) Impacts of exploratory drilling for oil and gas on the benthic environment of Georges Bank. Marine Environmental Research 27: 77–114.

82. Norkko J, Norkko A, Thrush SF, Valanko S, Suurkuukka H (2010) Conditional responses to increasing scales of disturbance, and potential implications for threshold dynamics in soft-sediment communities. Marine Ecology Progress Series 413: 253–266.

83. Hannah RW, Jones SA, Miller W, Knight JS (2010) Effects of trawling for ocean shrimp (Pandalus jordani) on macroinvertebrate abundance and diversity at four sites near Nehalem Bank, Oregon. Fishery Bulletin 108: 30–38.

84. McBride GB (1999) Equivalence tests can enhance environmental science and management. Australian & New Zealand Journal of Statistics 41: 19–29.

85. 84. Archambault P, Banwell K, Underwood AJ (2001) Temporal variation in the structure of intertidal assemblages following the removal of sewage. Marine Ecology Progress Series 222: 51–62.

86. Jones DOB, Gates AR, Lausen B (2012) Recovery of deep-water megafaunal assemblages from hydrocarbon drilling disturbance in the Faroe-Shetland Channel. Marine Ecology Progress Series 461: 71–82.

87. 86. Dernie KM, Kaiser MJ, Richardson EA, Warwick RM (2003) Recovery of soft sediment communities and habitats following physical disturbance. Journal of Experimental Marine Biology and Ecology 285–286: 415–434.

Citations

CHAPTER 1

Thanes Weerasiri, Wanpen Wirojanagud, and Thares Srisatit, "Assessment of Potential Location of High Arsenic Contamination Using Fuzzy Overlay and Spatial Anisotropy Approach in Iron Mine Surrounding Area,"The Scientific World Journal, vol. 2014, Article ID 905362, 11 pages, 2014. doi:10.1155/2014/905362.

CHAPTER 2

Zheng Shen, Frederick E. Beck, and Kegang Ling, "The Mechanism of Wellbore Weakening in Worn Casing-Cement-Formation System,"

Journal of Petroleum Engineering, vol. 2014, Article ID 126167, 8 pages, 2014. doi:10.1155/2014/126167.

CHAPTER 3

Wojciech Naworyta, Szymon Sypniowski, and Jörg Benndorf, "Planning for Reliable Coal Quality Delivery Considering Geological Variability: A Case Study in Polish Lignite Mining," Journal of Quality and Reliability Engineering, vol. 2015, Article ID 941879, 9 pages, 2015. doi:10.1155/2015/941879.

CHAPTER 4

B. Raju, B. Suresha, R. Swamy and B. Kanthraju, "Assessment of Cutting Parameters Influencing on Thrust Force and Torque during Drilling Particulate Filled Glass Fabric Reinforced Epoxy Composites," Journal of Minerals and Materials Characterization and Engineering, Vol. 1 No. 3, 2013, pp. 101-109. doi: 10.4236/jmmce.2013.13019.

CHAPTER 5

C. Kane, F. Fussi, M. Diène and D. Sarr, "Feasibility Study of Boreholes Hand Drilling in Senegal—Identification of Potentially Favorable Areas,"Journal of Water Resource and Protection, Vol. 5 No. 12, 2013, pp. 1219-1226. doi:10.4236/jwarp.2013.512130.

CHAPTER 6

K.A. Fattah, S.M. El-Katatney, A.A. Dahab, Potential implementation of underbalanced drilling technique in Egyptian oil fields, Journal of King Saud University - Engineering Sciences, Volume 23, Issue 1, January 2011, Pages 49-66, ISSN 1018-3639, http://dx.doi.org/10.1016/j.jksues.2010.02.001.

CHAPTER 7

Charles Fairhurst (2013). Fractures and Fracturing - Hydraulic fracturing in Jointed Rock, Effective and Sustainable Hydraulic Fracturing, Dr. Rob Jeffrey (Ed.), ISBN: 978-953-51-1137-5, InTech, DOI: 10.5772/56366.

CHAPTER 8

R. Gates,.Daniel O. B. Jones, Recovery Of Benthic Megafauna From Anthropogenic Disturbance At A Hydrocarbon Drilling Well (380 M Depth In The Norwegian Sea) DOI: 10.1371/journal.pone.0044114.

Index